John Gunther has written this memoir, which includes a chapter by Johnny's mother, Frances Gunther. The book is a skillful and loving evocation of a boy's mind and behavior, in which maturity and unusual intellect were mixed with the humor, excitement and moods of youth. In addition, the book is a remarkable account of the limitations of medical science in the treatment of a brain tumor. There were operations and X-rays, specialists from all over the continent, successive treatments —some bizarre, some painfully routine— periods of despair and a few heartbreaking days when it seemed, unbelievably, that all would be well — all this reflected in Johnny's spontaneous and sensitive reactions.

Death Be Not Proud is a re-creation of the boy who was Johnny Gunther. It is an inspiring story, and a moving account of doctors and medicine.

Neither publisher nor author is deriving any profit from the sale of this book because of their donations to cancer research for children.

DEATH BE NOT PROUD

A MEMOIR

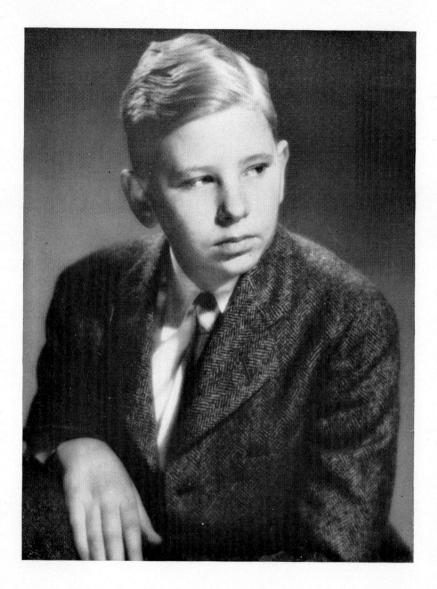

Johnny at 15

DEATH
BE NOT PROUD

A MEMOIR

By JOHN GUNTHER

NEW YORK
HARPER & BROTHERS
1949

DEATH BE NOT PROUD

Death, be not proud, though some have called thee
Mighty and dreadful, for thou art not so:
For those whom thou think'st thou dost overthrow
Die not, poor Death; not yet canst thou kill me.
From Rest and Sleep, which but thy picture be,
Much pleasure, then from thee much more must flow;
And soonest our best men with thee do go—
Rest of their bones and souls' delivery!
Thou'rt slave to fate, chance, kings, and desperate men,
And dost with poison, war, and sickness dwell;
And poppy or charms can make us sleep as well
And better than thy stroke. Why swell'st thou then?
    One short sleep past, we wake eternally,
      And Death shall be no more: Death, thou shalt die!

                        —JOHN DONNE

I want to acknowledge with the deepest thanks the assistance Frances Gunther has given me in preparation of this memoir. It could not have been written without her wise and discriminating help. In particular many anecdotes about Johnny and several of the lines of his dialogue come from her memories and records, which she has generously shared with me.

J.G.

CONTENTS

PART ONE

Foreword

HIS is not so much a memoir of Johnny in the conventional sense as the story of a long, courageous struggle between a child and Death. It is not about the happy early years except in this brief introduction, but about his illness. It is, in simple fact, the story of what happened to Johnny's brain. I write it because many children are afflicted by disease, though few ever have to endure what Johnny had, and perhaps they and their parents may derive some modicum of succor from the unflinching fortitude and detachment with which he rode through his ordeal to the end.

Johnny was conceived in California, carried across the bosom of the American continent and the Atlantic Ocean by his mother, and born in Paris, on November 4, 1929. We moved to Vienna when he was a few months old, and he went to kindergarten there and had splendid holidays in the Austrian Alps. We moved again to London when he was six, and he had a year and a half in England. Then we returned to the United States. Johnny went to the public school in Wilton, Connecticut, and to several other schools, including Lincoln in New York City, which he loved with all his heart, and finally to Deerfield Academy, in Deerfield, Massachusetts. He died on June 30, 1947, when he was seventeen, after an illness that lasted fifteen months. He would have entered Harvard last autumn had he lived.

I must try to give you a picture of him. He was a tall boy, almost as tall as I when he died, and skinny, though he had been plump as a youngster, and he was always worried about putting on weight. He was very blond, with hair the color of wheat out in the sun, large bright blue eyes, and the most beautiful hands I have ever seen. His legs were still tall hairy stalks without form, but his hands were mature and beautiful. Most people thought he was very good looking. Perhaps as a father I am prejudiced. Most people did not think of his looks, however; they thought of his humor, his charm and above all his brains. Also there was the matter of selflessness. Johnny was the only person I have ever met who, truly, never thought of himself first, or, for that matter, at all; his considerateness was so extreme as to be a fault.

There was that day after the first operation, the operation that lasted almost six hours, when Dr. Putnam thought it wise to tell him what he had. Johnny was too

bright to be forestalled by any more myths or euphemisms. As delicately as if he were handling one of his
own instruments of surgery, Putnam said quietly,
"Johnny, what we operated for was a brain tumor."

Nobody else was in the room, and Johnny looked
straight at him.

"Do my parents know this? How shall we break it
to them?"

Then, some months later, when he seemed to be
getting better, he felt the edge of bone next to the flap
in the skull wound, and looked questioningly and happily at the doctor—a different doctor—then attending
him. The doctor was pleased because the bone appeared
to be growing back, but with a crying lack of tact he
told Johnny, "Oh, yes . . . it's growing . . . but in the
wrong direction, the wrong way."

Johnny controlled himself and said nothing until the
doctor left the room. His face had gone white and he
was sick with sudden worry and harsh disappointment.
Then he murmured to me, "Better not tell Mother it's
growing wrong."

I do not want this brief foreword to be a Bright-
Sayings-of-the-Children essay or the kind of eulogy that
any fond and bereaved parent may be forgiven for trying
to put on paper. What I am trying to tell, however
fumblingly and inadequately, is the story of a gallant
fight for life, against the most hopeless odds, that should

convey a relevance, a message, a lesson perhaps, to anybody who has ever faced ill health. But to do this I must first stake out a few facts about Johnny and give the reader some detailed impression of his character. I want to make some part of him come alive again, if only in the feeble light of words. So let me go back briefly into the moist net of memory.

I have been rummaging this past month through all the papers and things he left, things we had saved and treasured from his earliest days—the notebook Frances kept recording his weight and height and other such memorabilia when he was an infant, the first drawings he made, the earliest snapshots, then evidences of his schoolwork, his letters, his report cards and themes and diaries. Here on my desk is a bookmark, bright with enameled colors, that his beloved Austrian governess helped him make when he could not have been more than two; here also his very last chemical formulae and mathematical computations and calculations, which are far beyond my lay comprehension.

Johnny's first explorations of the external world were in the form of pictures—graphic art. Some of his paintings still hang on the walls in the house in Madison, Connecticut. The violence with which a child sees nature! The brilliant savagery of the struggles already precipitated in an infant's mind! Here are tigers of the most menacing ferocity—bloody, chewing up baby

[6]

lambs, with their red jaws open and foaming; then by contrast a group of somnolent, placid, herbaceous elephants; then landscapes mostly in green, with blue trails leading to jagged mountains, and a dirigible soaring in the sky; then boats dancing on blue water, boats with multitudinous white shining sails. Later came airplanes, locomotives, heavy trains. Johnny always had an acute interest in transportation. One of his earliest concepts was "Smoky," a magic personification of a machine that conquered all frontiers of time and space.

Music came next. Not all children are Mozarts; but almost all are geniuses at one thing or another before they are ten. Johnny had a considerable musical talent, though he did not push it far. He was, like the youngster in Aldous Huxley's beautiful story *Young Archimedes*, more interested in the structure of music, in its mathematics, than in playing tunes or listening to melodies. He took violin lessons early and kept his precious recorder close to him till the day he died. When he was about ten he became fascinated with the woodwinds, especially the bassoon. For years his ambition was to own and play a bassoon; this instrument, however, does not exist in miniature form, and of course he was never able to handle one. But he saved regularly out of his allowance to buy one later. He would sit by the hour listening to woodwind music, and we bought him

[7]

practically every record that exists in which the bassoon is prominent. I remember coming home one day with the Schubert Octet, Opus 166. He listened to it, rocking with excited glee, "Oh, boy! Oh, boy!" Next year he was hard at work composing; in April, 1940, he finished what he called the rondo of his first "symphony." Later that year he played several of his own compositions at a Lincoln School recital. I was doing a job somewhere and called him on the phone to say that, unhappily, I could not get back to New York in time to hear him. He replied drily, "You could certainly get here if you hired an airplane."

Concurrently came an engrossed interest in various games, particularly chess. He was beating me at chess easily by the time he was twelve. He became fascinated, too, in weather forecasting, and built up a formidable array of charts and instruments; like Frances, he loved weather in the abstract, all and any kinds of weather. He loved gardening and he tried experiment after experiment in gardening without soil, using hydroponics. He loved puppies and small cats and turtles. He loved to collect rocks and study them and to smelt bits of iron out of ore. He loved magic and card tricks, and worked out dozens of tricks, some of which I never got onto. Once we dined with Cass Canfield, my publisher, and Johnny went through his repertoire. "Well," Cass said, "all I hope is that he grows up honest."

I do not mean to give the impression that Johnny was any prodigy. He was good at some things, not good at others. His I.Q. at one of his schools was, we were told, the highest ever recorded there, way above the genius level, but his marks were often indifferent. He was a great procrastinator. In one examination, I remember, he was the best of his class in content, and the worst in neatness. He lacked assertiveness and self-reliance. He had plenty of stick-to-it-iveness, but this was often badly focused, and he was nowhere near so efficient as he might have been. But in the last quiz he ever took, he got a 99. Twice he won the current-events test at Deerfield, and—characteristically—never told us so. When he arrived at Deerfield he insisted on taking five subjects instead of the usual four. This is a stiff school, and Mr. Boyden, the headmaster, told us afterward that he had never dreamed that Johnny would be able to keep up with all five. The fifth subject was geology; Johnny insisted on taking it because he was so determinedly eager to have a science course every year. He got straight A's in the final examinations in both algebra and geology. Not till a long time later, when I happened to run into a friend whose son was also a Deerfield boy, did I learn that no one in the school's entire history had ever taken five subjects in a single year before.

Already the major line of his brief life was drawn

sharp and clear—his passionate love for science. Many things Johnny did sloppily, and after many false starts and delay, but his scientific drawings and charts almost always had an exquisite precision. By the time he was fifteen he had veered away from applied science; what he loved and intended to devote his life to was scientific theory, science in the abstract, as pure and undiluted as he could get it. He had not quite decided whether to be a physicist or a chemist; probably he would have chosen the trans-Uranian borderland between the two. He was an experimentalist. He liked the pragmatic approach. He had a small laboratory in our apartment; I think now of the procession of happy hours he spent there with his chemicals, weights and measures, retorts, tools and electrical apparatus. Once a lady joked with him about his interest in sums and figures. He replied, "You don't understand at all. Arithmetic bores me. What I am interested in is mathematics."

Atomic physics fascinated him, of course. As the prize in one of the quizzes he won, he chose an advanced college text on the atom—something far beyond his powers at the time. He would secretly read it at night, a little at a time, absorbing it "by osmosis" as he said, after the dormitory had gone to bed. In February, 1945, seven months before the new world blew wide open at Los Alamos and Hiroshima, he wrote a theme actually about atomic fission and what its consequences might well be. The last thing I have in his handwriting is a pitiful distorted scrawl—and once his writing had been so graceful and confident!—

Wonderfull! (sic) Relativity of dimensions.

* * *

Perhaps I might allude to other aspects of his personality, though I hope these will become clear, in outline at least, in the pages to follow. Johnny was a sensitive boy, quite diffident and shy, quite hesitant, a boy who chose his friends carefully and then held them, and very serious—though he loved to laugh and his smile was radiant. Perhaps I have given the impression that he was somewhat on the stuffy side. Never! In plain fact he was the sunniest of creatures.

He adored fantasy. He was an inveterate daydreamer and once when Frances mentioned his habit of introspection he replied, "Oh, but I spend hours analyzing myself!" One of his headmasters, Mr. Hackett of Riverdale, who was very fond of him, told us when Johnny was about twelve, "You know your son lives in a world altogether his own."

This is from a letter a friend wrote Frances after his death:

I am always a little afraid of children—and yet when I first met Johnny as a little boy, he completely wiped out that gulf which I usually feel with the younger generation.

I never felt it with him. He was always a person and my friend.

Another friend began a long letter:

He was such a mixture of both of you, and so himself. He was so young and yet he took such a serious view of life. So many of the children of my friends are inconsiderate brats that I was almost taken aback by his quiet, attentive manners. How almost formal he was.

When he read a book, he would often stop, as if considering what he had read, then go back and read something over. He talked very little about his school but he never seemed to have that competitive brash something so many schoolboys have.

Another of our friends told us later that, dropping in to see us when Johnny was home, he was careful not to make any unwise remark—he always wanted to be at his best with him, fearing, too, that his best might not be quite good enough, since Johnny, even at this early age, reminded him "of a Chinese sage."

Johnny was irritated by his lack of prowess at athletics. But you should have seen him sail a boat! He was never one of a gang, and had little interest in team play, but he swam very well, rode, and played lacrosse. At Camp Treetops in the Adirondacks, where he spent several summers, he was captain of the baseball team—if only because everybody liked his humor and trusted his fairness of mind. One of his camp counselors wrote:

His swimming is excellent, and he is acknowledged by the boys themselves to be the best long-distance swimmer in the group. . . . It's his nature to be jolly and unaggressive. . . . If he can't hit a home run, he's not the least unhappy about it. . . . By every rating I have made out, he was the most popular boy in the group.

There was also a note:

We noted that John never arrived on time for any distasteful task.

"It is extremely difficult," wrote one of his early teachers, "for Johnny to keep his personal belongings in order." Johnny's slouch! The way his hands dug deep into those long pockets! The way he hesitated and groped and said "Um" and "Um"! He had an almost unparalleled capacity to lose things. Once at Deerfield he mislaid and never found one of a pair of shoes. Once we walked in a windy snowstorm from our apartment to a restaurant around the corner, and in that distance of two hundred yards he managed somehow to lose his hat. I was with him; we searched everywhere; the hat had simply disappeared, and was lost without a trace.

"I keep working on his absent-minded ways," another teacher reported. "His books and possessions are left behind him everywhere. Life is entirely too full for him to manage the practical details of living."

I cannot convey Johnny's originality, his wit, his use of language, without searching further back. Once Frances and I drove him out to Oyster Bay for a Sunday luncheon party—Johnny was about ten, and this was probably his first grown-up luncheon. He expressed some curiosity about which breed of Roosevelts we were visiting, and I tried to sort the family out for him. He said, "The gist of it is—are they Roosevelts who would be for or against the WPA?" At lunch he saw Dr. Lin Yutang, the first Chinese he had ever met, and straightway went up to him, asking, "Is it true what my father says, that no Chinese ever eat cheese?" Dr. Lin ruined my authority as a parent by walking firmly to the buffet and putting a large piece of cheese in his mouth. On the way home we started to talk about the transmigration of souls and what various people might have been in former reincarnations. Frances said that she was quite sure that she had once been some kind of fish. Johnny thought for a moment and then said that he knew, at least, what he wanted to be in some future avatar—a sperm whale.

Like most sensitive youngsters, Johnny took a good deal of time to cross the shadowy frontier between childhood and boyhood; one could never be quite sure how adult his reactions were going to be. Once he said to a kind lady who was attempting to interest him in religion, "Of all the times that I am not interested in Christianity, this is the time I am least interested." At about the same time, he announced that during his spare time at school (spare time!—he was editor of the school paper then and studying the violin as well as doing his class work) he was opening a tie-pressing business. This was to earn money in case I should be drafted! Once he asked why so many artists, painters, and musicians were a little crazy. We explained that some people thought that artists were crazy simply because they were in advance of their time. Johnny paused a moment, and then delivered himself of the following: "The only crazy thing I do is that when I reach for my slippers under the bed, I put them on upside down, to my great dismay!"

His life was packed with everything from postage stamps to mineralogy to electric trains to how to cook. He announced to me firmly, "I have too many hobbies, and I am going to give up five or six of them!" As a self-imposed task, he set about classifying and cataloguing our gramophone records, and he went at this with the utmost attention to minute detail. Composers were indexed on a card of one color, executants on another, titles on another, and so on. He typed each card patiently, sometimes with six or seven different entries for a single record. One day I protested that an index or catalogue was a means to an end, not an end in itself, and that he shouldn't get bogged down in such minutiae. His reply was, "Don't you understand that I have become a perfectionist, just like you!"

This memoir will contain a good many of Johnny's

jokes, asides, and wisecracks. Actually these were a comparatively recent development. For a long time he was too shy to talk much in company; he was almost painfully ill at ease. Then, in the year before his malady struck him down, his confidence grew steeply. He conquered his own shyness; it was quite a battle, and the dry, thoughtful humor that came to characterize him was among other things a mark of deliberate endeavor. Even as a child, though, when he did speak, what he said was often worth listening to. He could not have been more than six when he announced to me that he had discovered what God was. "God is what's good in me."

When he was about thirteen he had a talk with a doctor who asked him to list the things he "wished" most. His answers were, first, happiness and, second, to do some good for the world. Then he added a third— an extra week's vacation!

The doctor asked him what he would like most if a magician could change everything by the whisk of a wand. Johnny replied with sharp knowledge of his own defects: (1) To be a bit more talkative. (2) To be a better athlete. (3) To stop putting things off. (4) To be tidier.

Finally to the question "What three things bother you most?" Johnny made an omnibus answer: "School problems, studies, sports, and relations to other boys."

He added, "Relations to parents" and then changed his mind: "I do not think this bothers me."

*   *   *

Johnny and Frances were indeed marvelously close. I was close too, closer I think than most busy fathers with growing sons, but not in the way Frances was. Also Johnny saw much more of us together than a child usually sees of divorced parents. Our modus operandi was that he spent the winter and spring holidays with me in New York, and the summers with Frances in Connecticut. But I saw him frequently in the summer too, and almost always for a week or so at the beginning and end of each summer, and Frances similarly saw him a great deal in New York in winter. We overlapped all the time. For years though I traveled widely I specifically arranged almost every job and itinerary so as to be with him during his holidays.

From Frances he got a tremendous lot—his gift for fantasy, the realism and long view of his intelligence, his delicacy of perception, his creative curiosity. From the earliest days she saved out for him books and poems, pictures and puzzles and jokes and parables and anything that would enlarge his conception of life as something to be truly lived. For year after year she encouraged him to read, to think for himself, and to admire wisdom, truth, and beauty.

[17]

When he went up to Deerfield for what we did not know would be the last time, she wrote him:

> Love God
> Even if you get four A's, be humble.

* * *

Overnight, it seemed, Johnny became startlingly, almost frighteningly, intellectually mature. We were walking back from the theater and passed a newsstand and I asked him if he wanted the next day's paper. He said, "No—sufficient unto the day is the evil thereof." One evening we asked him if he'd like to see some movies taken of him when he was a child. He replied, "Only if they're not too recent—the past is tolerable if remote enough."

Presently he became a good critic of whatever I happened to be writing at the time. I showed him the first part of an unsuccessful long story; he read it, puzzled, and then exclaimed, "But when does something happen?" He had gone straight to the heart of the difficulty. One of the earliest times we ever talked as man to man about a nonpersonal thing came late in 1943, when he had just finished a course in English history and I was driving him in from Connecticut. The subject of the conversation was William Ewart Gladstone.

Quite parenthetically—this goes back to his early

[18]

childhood—it was a fragment of his conversation that helped set me out on ten years of work. Inside Europe appeared when we were still in London, and he noticed a shop window, Selfridge's I think, full of the English edition of that book. At this time it had never occurred to me to follow Inside Europe with similar books. But Johnny, aged about six, looked again at the window and declared, "I suppose you will be doing Inside Australia next—or Inside the North Pole."

He was never a great reader—science interested him much more. But as he developed he dipped into books occasionally and I remember how fresh his comments were. Once I asked him what I could bring him to read, and he answered, "Anything provided it is not by George Eliot." Carl Sandburg's Lincoln impressed him greatly; in fact, he used his free time during the whole of his first year at Deerfield to read this huge book entire. Incidentally he always subscribed to the New York Times at Deerfield, to keep abreast of the news, though this took a sizable chunk out of his allowance. Frances showed him an advertisement of Upton Sinclair's Lanny Budd saga when he was in hospital. His simple comment was, "I prefer to take my Superman straight."

When he was home from school and I was broadcasting I would often take him to the studio. On November 7, 1942, I finished writing a long broadcast, mostly about Russia, at 7 P.M. Then a story began coming in on the ticker about fleet movements in the Atlantic and Mediterranean. So I did a new lead paragraph and went home. Johnny was waiting for me there. I listened to the 8:55 news, which I rarely do, but there was nothing new. Then at 9:02 my secretary called to say that we were landing troops in North Africa. I went down to the office right away and took Johnny with me. I told him what had happened and that of course I would have to rewrite the entire broadcast and maybe would have to ad lib most of it. Instantly, thirteen years old as he was, he asked just two questions: what would the French fleet do, and what about De Gaulle. Neither of these two angles had occurred to me, but it was obvious as soon as Johnny offered the words that they would dominate all that happened next.

His intelligence was, above all, detached and reasonable—and what is a mind for, except to reason with? He could be crisp and sharp-eyed even about phenomena very close. For instance about one of his schools he said, "I would make the following criticisms. First, too much attention to marks. Second, too much religion. Third, no time for me to develop my own interests. Fourth, group discipline may be imposed unfairly." He added that something must be wrong with the entire educational system in that, though he was three years behind in French, he made up the whole gap in one term, but had to remain now with the class he caught up with.

Johnny's generosity—he would give anything away—his affectionateness and amiability, concealed a considerable sturdiness of character. Everybody loved him —down to the corner cop. I have never known anyone so loved. This led some people, especially when he was very young, to baby him, which he resented in his calm, reasonable way. When he was at Riverdale one teacher wrote in her annual report that she had never known him to raise his voice, even out of class. Yet, a couple of years later (I did not discover this till much later) he took off fifteen pounds by limiting himself to a diet of 1,200 calories a day—what an effort of will on the part of a hungry youngster!—without telling a soul. He was gentle, yes, but he did not lack will, as I hope this chronicle will show.

A footnote as to food: once I took him to a restaurant that served a particularly luscious form of ice-cream cake. He nibbled at it tentatively and then his will power broke down and he ate like a trencherman. "Papa," he said, wiping his mouth, "that was simply irresistible."

There was absolutely no trace in him of malice, acquisitiveness, or vulgarity. It was almost impossible to make him lose his temper, and during the whole course of his illness and the frightful demands it made on him, I scarcely ever heard him utter a serious word of protest or complaint. He was always obedient—if only because he wanted so desperately to get well. Johnny was a careful boy; he was truly modest and he believed in the goodness of people—he gave out something, so that people entering a room felt at once a warm, compelling, sunny contact. But he was, like most exceptional children, a very complex character. He had the dominant qualities of his generation—factuality, understatement. He was not a backslapper and he was hopeless at hail-fellow-well-met contacts, but everybody respected him. His judgments were quite simple: he disliked bad things, and liked things good. Johnny was as sinless as a sunset. When he uttered that childhood wish—"to do some good for the world"—he was reflecting all the gifts that had been given him, of goodness, gentleness, and warmth of spirit; he was one of those who thought earnestly that he owed the world a living, not vice versa. But he never got a chance, and the world is much the poorer for it.

But it is time now to go on to what I have to tell.

DEATH BE NOT PROUD

A MEMOIR

Death Be Not Proud

CHAPTER 1

JOHNNY came home for the Christmas holiday in 1945, and he looked fit and fine. He
was lengthening out physically and otherwise, as children do all of a sudden, responding as
it were to the release of some hidden inner spring.
We saw a lot of each other, and just before getting
on the train to return to school in January, he exclaimed, "Pop, that was the best ten days I ever
had!" He didn't often confess personal emotions so
freely, and I was pleased. Then in March, 1946, he
came down again for the long spring holiday.
Frances and I took him to several Broadway shows,
including *Show Boat* and *Antigone*—he liked *Antigone* best; he went to lectures on atomic physics;
Frances took him to the public dinner given by the
City of New York to Winston Churchill—it was
the first, and last, time he ever wore a dinner jacket,

borrowed from his uncle; he won the critical game in a chess match against another school (he was captain now of the Deerfield chess team); he monkeyed with his chemicals and read the manuscript of the early chapters of *Inside U.S.A.* which was just then getting under way. I thought he seemed tired, but I did not take this seriously, believing it to be the normal reaction from a regime as vigorous as that of Deerfield, together with the strains of adolescence. He had his usual check with Traeger, our family physician, who pronounced him perfectly all right. Also he had a check with an eye doctor. This was important. Johnny had suffered some eye strain the summer before and was taking exercises to strengthen his visual acuity. The eye doctor found nothing wrong; in fact, the eyes had improved to a considerable degree. The day after the examination by Traeger, Johnny complained suddenly of a slight stiff neck. If this had happened before Traeger saw him, I would have been more concerned, but since he had just been given a clean bill of health, we did not take anything so minor as a stiff neck seriously. Indeed, it disappeared after a day, and Johnny went back to school, sighing a little that the holiday was over but happy and full of energy and anticipation.

Deerfield had an infantile paralysis case that spring, and, as is the custom of the school with its strict standards, all parents were notified at once. Then in the third week of April I had a wire from the school doctor, Johnson, saying that Johnny was in the infirmary but, though he had a stiff neck, there was no indication of polio and we were not to worry. Nothing at all alarming was indicated. Boys get stiff necks and Charley horses all the time. In fact, Dr. Johnson said, he was informing us of Johnny's complaint only because, knowing of the polio scare and hearing that he was in the infirmary, we might think that he did have polio, which he didn't. I called Johnny up, and we talked briefly. He was lonely, and fretful at missing a week of class work, but otherwise nothing seemed to be amiss. He was going into the nearby town the next day to have a basal metabolism test, and Dr. Johnson asked me to find out from Traeger when he had last had a basal, and what it was. I reported all this to Frances, and thought little more of it. Later we found that Johnny might not have gone to the infirmary at all, since he would never admit it when he was ill and never complained, except that one of his classmates, observing his stiff neck, insisted on his seeing the doctor. Then, wisely, Dr. Johnson held him for observation. Had this not happened, he might have died then and there.

At about three in the afternoon on Thursday, April 25, the telephone rang in our New York apartment. Just at that moment I had finished the California chapters of my book, and I had intended calling Johnny that night to tell him.

Without hesitation or warning Dr. Johnson said, "We've had a doctor in from Springfield to see your son—Dr. Hahn, a neurologist. Here he is."

Dr. Hahn said, "I think your child has a brain tumor."

I was too stunned to make sense. "But that's very serious, isn't it?" I exclaimed.

Dr. Hahn said, "I should say it *is* serious!" He went on, in a voice so emphatic that it was almost strident, "His disks are completely choked."

"His what?"

"Ask any doctor in the world what that means— choked disks!" he shouted.

He proceeded to describe other symptoms, and implored me with the utmost urgency to get in touch at once with Dr. Tracy Putnam, the best

[28]

man for this kind of thing anywhere within range; in fact, even before talking to me, he and Johnson on their own responsibility had put in a call for Putnam. The next half hour passed in a grinding crisscross of calls. I talked to Traeger, I called Deerfield back, I got in touch with Frances who was out in Madison, I reached Putnam, I consulted Traeger once more, and by half-past four I was at 168th Street, waiting in Putnam's office. We picked Frances up in New Haven and, driving hard through greasy rain on an ugly, gritty night, with the windshield smeared all the time by fog and thick penetrating mist, reached Deerfield at about ten. Putnam said little as we drove, with our hearts dropping out of us. Five minutes after I got there I knew Johnny was going to die.

I cannot explain this except by saying that I saw it on the faces of the three doctors, particularly Hahn's. I never met this good doctor again, but I will never forget the way he kept his face averted while he talked, then another glimpse of his blank averted face as he said goodbye, dark with all that he was sparing us, all that he knew would happen to Johnny, and that I didn't know and Frances didn't know and that neither of us should know for as long as possible.

Johnny himself was cheerful. They hadn't told him that we were en route; he jumped up in bed as we came in and murmured, "Well, for goodness' sake!" I saw that his right eye hung down slightly on his cheek.

Johnny thought that he had polio. He tried to grin.

Frances stayed with him; I talked to the doctors, and later we told her that Johnny had some sort of "pressure" (that was the euphemism the doctors chose to employ at this time) within his skull, and must be operated upon as soon as possible. She took this without flinching, and returned to Johnny to make him comfortable. I could not bear to look at his eye, limp on the cheek that way. Then I got a more or less consecutive account of what had happened.

Dr. Johnson, suspecting nothing so grave as this, had sent Johnny into Greenfield that morning—luckily with a nurse—for his metabolism test. Johnny talked of playing hooky and going to the movies. Then the nurse—good girl!—noticed that he seemed to be stumbling ever so slightly and that when he walked down a corridor or through a door, he brushed the wall on the left side. On closer examination she saw that his eyes were not co-ordinating; he had a squint. This scared her, she reported the matter to Johnson, Johnson made an optical check and at once called Hahn in, and Hahn took a spinal tap which showed an almost unbelievably heavy pressure in the fluid encasing the brain, as well as the choked optic disks that indicated fierce pressure behind the optic nerve.

That first spinal tap!—the first of many, and spinal taps can be frightening as well as painful. All the other tests!—some of them lasted a full hour, with every reflex in the body being labori-ously investigated. For fifteen months, hardly a week passed that Johnny did not have some kind of examination or other; for month after agoniz-ing month there were the bandages and dressings to be changed every day; day after day, with never even the slightest respite, he faced and went through the most exhausting medical procedures. Yet, I give my word on it, no whimper ever came out of Johnny after the first operation, no word of unreasonable protest or appeal, no slightest con-cession to terror or giving way to misery. In fact, his reactions were by and large the opposite. He got angry sometimes but he never blubbered. Soon he became fascinated, as a bright child would, with his own illness, and particularly with all the techniques the doctors applied—he demanded to know the precise reason for every step both theoretically and in the realm of concrete therapy. He helped the doctors in the most active possible manner and indeed came presently to regard himself with detachment and curiosity almost as another person on whom emergency experiments were being performed. "I am quite a guinea pig," he would say.

And the doctors! So many doctors! We had thirty-two or thirty-three—maybe more, including some of the most famous specialists in the world—before the end. Every doctor who dealt with him, except possibly one, every one of those thirty-two or thirty-three doctors came to love him, and I truly think half a dozen would have gladly given their lives to save him. But very soon we discovered several things about doctors. One is that they seldom, if ever, tell you everything. Another is that there is much, even within the confines of a splinter-thin specialty, that they themselves do not know. Let me salute all those doctors. They had the best will in the world, and nothing in the entire province of modern knowledge applied to this particular ailment was left unsearched; indeed we tried some things that had never been tried before, but the frontiers of medicine are in some fields astonishingly limited—not to say unknown —and there are still mysteries in Johnny's case that no medical man can altogether account for.

Of all the doctors, the chief was this gentle and sensitive man whom we drove to Deerfield that first evening, Tracy Putnam. I had no idea then of his eminence. I didn't even know he was a surgeon. As of that time—he has since moved to California— he was Professor of Neurology and Neurological Surgery at Columbia University and Director of the Neurological Institute of Columbia-Presbyterian Hospital, commonly known in New York as Medical Center. That a man of this rank should leave his desk and go to a child's bedside hundreds of miles away on five minutes' notice is sufficient indication of his character. Probably Putnam is the most delicately fastidious expert in his field I have ever met. He has a great quality of sensitive reserve also; brain surgeons seldom give themselves away. Johnny used to say—or perhaps it was Frances—that he resembled Buddha.

Again that scene—the white frame building with the tall Deerfield elms outside, beaten by a howling wind; Johnny's small room with the bed stuck out sideways from the wall because there was so little space; the lights dim as it became midnight, and the doctors tiptoeing and the nurses whispering; the first talk the doctors had when they wouldn't let Frances or me into the room, and how long it lasted, while an unbelieving sickness, a bewildered stupefaction, rose in our protesting hearts; Johnny's own dazed smile and one murmured sentence, "I know it can't be really serious or they would have taken me to a hospital."

Putnam ordered Johnny to be brought into New York by ambulance; we made the arrangements, and set out early the next morning. To keep Johnny warm as we lifted him into the ambulance, the nurse pulled a gray blanket over his face. Frances helped her. I did not want to watch. It was a long ride in the cold, sullen, slippery rain. Frances held Johnny's hand while he dozed. The Neurological Institute rises stiff and tawny near the Hudson just below the silvery spindles of the

George Washington Bridge. That building!—it became the citadel of all our hopes and fears for more than a year, the prison of all our dreams. A comfortable room with a broad view of the river was ready, Johnny was transferred gently to a bed, and we found ourselves sucked at once into the vast mechanism of a modern hospital, with all its arbitrary and rectilinear confusion.

The next morning Johnny was well enough to ask me how much the ambulance had cost.

I told him, and he replied, "A gyp."

The nurse asked him if he had had a bowel movement the day before. He replied, "Nominal."

His eye looked better. It did not have that dreadful droop. But later that day he developed an excruciating headache, the only fierce and intense pain he suffered during the whole course of his illness—a small mercy, perhaps, but one to be devoutly grateful for. The brain controls pain in other parts of the body, but there are no sensory nerves in brain tissue itself; you could cut a person's brain apart bit by bit, and there would be no pain. What causes headache is swelling or inflammation of the membranes surrounding the brain, or pressure on the tissue from a foreign mass; this is what happened that day and Johnny muttered angrily about the savage pain and tried to analyze it. "Pop, I feel a sword go through my head at every pulse beat." The usual painkillers were forbidden, because they might interfere with the tests remorselessly going on. Finally, an injection of caffeine relieved him somewhat, and he had some medicaments by mouth. He asked the doctor for the chemical formula of one that, he said, "was, of all the concentrated essence of bitterness ever invented, the bitterest." It was something quite simple chemically—and he was disappointed!

A great number of intricate tests were necessary, including X-rays, an electroencephalogram, and visual field tests, all exhausting in the extreme, but necessary so that the tumor might be located as accurately as possible. The surgeon, when he went through the bowl of skull, wanted to hit the exact right spot. One of these tests, the ventriculogram first used successfully by the late Dr. Dandy of Johns Hopkins, actually entails drilling holes through the skull—of course it can only take place on the operating table before the actual operation. Meantime at least five doctors, all neurosurgeons, asked us questions. Any record of a blow? Any propulsive vomiting? Any chills or tremors? Any double vision, headaches, abnormal involuntary movements, dizziness, or disturbances in gait, taste, smell, or hearing? We answered, horrified, "No . . . No." This vicious invader had given us practically no warning. Several of the doctors were Latin Americans; their English was imperfect, and it was a trial for Johnny to answer so patiently what they asked, and then be unsure that they understood the answers. Above all, what he suffered from was lack of water. He was allowed only a bare minimum of fluid, since dehydration would tend to decrease the pressure inside his head. Came more and other tests. Johnny said wearily, "All this red tape—why can't those doctors get together?"

After the violent headache the first day, the only thing that really hurt him was the haircut when Tony, the barber, shaved his skull the morning of the operation. This can be very painful when the razor scrapes against the grain. Johnny gave out a fierce "Ouch!" and grabbed for my hand. Then he asked how he could go back to Deerfield inasmuch as crew haircuts are forbidden there. He looked at his shaven skull. "Papa, they aren't going to electrocute me, are they?" He tried to laugh, but his voice was a nervous little giggle.

Johnny's operation—this first operation—took

place on Monday, April 29, 1946. He went upstairs at 11:10 A.M. and came down at 5:20 P.M. Brain operations take an eternity because of the laborious procedure necessary. One of the doctors told me that its effect on Johnny would be approximately that of the explosion of a .45-caliber bullet against the head. Those six hours were the longest Frances and I ever spent. A couple of nurses asked us with the deadly casualness that nurses have, "Is he your only child?"

Between the Friday and the Monday I had tried to find out something about brain tumors. I prodded through several texts full of the frightful jargon of medical writers, and consulted Traeger almost hour by hour. Let me pay tribute now to the steadfastness of this splendid physician, with his tough, cheerful, realistic mind, and his intense devotion to Johnny, who acted as a kind of chief of staff for all the other doctors during the entire illness. A tumor is a growth. What I asked about first of all was whether or not Johnny had cancer. All cancers are tumors, but not all tumors are cancers by any means. By one definition, Johnny did have cancer; by another, he did not. That is, a brain tumor (though it may strike the spinal cord in some

rare cases) never metastasizes, i.e., spreads through
the body to attack liver or bones or what not. It is
not like a tumor of the breasts that may become a
tumor of the lung. But if malignant, it will spread
within the cranium itself like a spot in an apple
until the brain is destroyed. Therefore, it must be
removed at once. Also, inasmuch as the brain rests
within the hard shell of skull, there is no room for
expansion; if a foreign growth is present, the skull
itself must be opened or death will be caused by
pressure. The only other accepted means of therapy
for brain tumors are X-ray and, rarely, radium.
Now, to open the skull and extract all or part of a
brain tumor is a refinement of surgery of the most
elaborate kind. The operation, in fact, did not exist as a practical possibility until the advent of the
late Dr. Harvey Cushing. Almost all the great contemporary neurosurgeons—like Putnam himself
and Wilder Penfield—are Cushing men, Cushing
trained.

The technique (of course I am oversimplifying
vastly) is to locate the tumor with exactness, open
the skull, and remove as much of the tumor as
possible by suction and other devices, meantime
preventing hemorrhage by various means and so

keeping the patient alive. In blunt fact the operation itself, though prolonged, may not be violently
dangerous, and though the technical preparation
may be difficult, the actual surgery is of the most
primitive type—simple extrication of an alien
mass. Everything depends on the type of tumor the
surgeon may discover, and there are half a hundred
different kinds, some comparatively benignant,
some malignant in varying degree. The location
of the tumor is also of prime importance. Obviously a tumor close to the surface, in the so-called
silent areas of the brain, will be easier to extract
than one deep down. At first it was thought that
Johnny's tumor was of the fourth ventricle. When,
up in Deerfield, I asked Dr. Johnson what kind of
operation this entailed, he simply shook his head
and replied, "There have been very few successful
operations in that area." Actually Johnny's tumor
turned out to be near the surface and in the right
occipital parietal lobe, which made the procedure
somewhat less formidable, though, God knows, formidable enough. Another difference among tumors
is that some have a greater velocity of growth
than others and a tendency to recur. It was a bad
sign that Johnny's had apparently developed with

such speed. Another difference is that some are encapsulated, and so can be lifted out in a piece, like a marble stuck in jelly. These are comparatively easy to remove. But others are of an infiltering spidery type that creep and burrow along the minute crevasses of the brain, slowly but inevitably destroying function, and almost impossible to remove. If the surgeon goes too deep, the patient dies of loss of blood or, worse, so much healthy brain tissue has to be destroyed that he will be better dead.

Traeger gave up his practice for a day, no small sacrifice for a busy doctor, when Putnam asked him to attend the operation. We hoped desperately for good news to the last. Putnam had explained that he would know little until he actually went in. For all anybody really knew, Johnny might not have a tumor at all. What caused the pressure might be a blood clot. It might be a mere cyst. It might, even if a tumor, be of the most innocent kind. At about 4:30 that afternoon, while Putnam was still washing, Traeger (who had stood up hour after hour during the entire operation) came down and found us in the solarium near Johnny's room. I took one look at his face, and knew the worst.

Traeger had aged five years in those five hours. He was as gray and seared as if drawn by Blake. He could hardly control his features. Nor was I controlling mine. I took him aside and asked him just one question. "Was it encapsulated?" He answered, "No."

Putnam came down a few minutes later, briskly but looking like officers I have seen after a battle. I heard him call, "Where are the parents?" He walked me down the hall after a word of encouragement to Frances. "It was about the size of an orange. I got half of it."

What, in all sanity and conscience, *is* a brain? How in the name of heaven or all that is reasonable could an evil thing the size of an orange have existed in Johnny's head without making him sicker? The answer is, I daresay, suggested by the fact that immediately after the operation, he became tolerably well, even though half the tumor was still there. Part of the brain is nonsensitive tissue, and Johnny's tumor lay in a comparatively inactive region.

There was a bustle at the elevator and Johnny's bed was wheeled in. Frances, who was magnificently composed, took a brief walk. I was sick with

fright when I saw the oxygen and all the paraphernalia for transfusion. Already Johnny had had
a couple of pints of blood, we were told. But this
was routine, more or less. The doctors said that he
would be unconscious at least till the next morning, and that we might as well go home. I elected
to stay. Soon I got my first look at Johnny's face,
sideways on the pillow with a huge turban of bandage marked THIS SIDE UP. I retreated in dumb
shock. Both his eyes were stuck closed and he
looked as if he had two enormous shiners; his
whole face was the size and almost the color of a
football. But this was nothing to worry about,
since it was the result of edema, swelling, following the shock of operation—so I was told. Then I
noticed the emergency instruments on the bed
table and learned that a needle (it stayed there for
some days) was actually taped into the vein in his
arm, in case an emergency transfusion should be
necessary. This, too, was routine—or so I heard.
Putnam, in a raincoat, just preparing to leave the
hospital, came up at a run, after a nurse called
him. An injection of some stimulant was necessary.
Johnny was still in shock. Another doctor said
airily, "Oh, he'll last the night all right."

At about nine that evening, only a few hours after the operation, Johnny gasped and stirred, making a weak groping gesture with his enormously swollen mouth.

"Spit it out," the nurse said.

He replied in perfectly understandable words, "You know perfectly well I can't spit. I'm completely dehydrated."

The nurse stared at him dumfounded.

He asked then for Dr. Miller, one of Putnam's assistants, with whom he had been talking about chemistry just before the operation. I told Miller that Johnny had asked for him, and Miller could scarcely believe that a child who had just gone through such an ordeal could be capable of speech.

Johnny recognized me after a while and whispered, "Hello, Pop." Pause. "Are there going to be any more tests?"

"Good Lord, no! You're all through with tests. Don't you realize you've had quite a serious operation?"

"Of course," Johnny answered. "I heard them drilling three holes through my skull, also the sound of my brains sloshing around. From the sound, one of the drills must have had a three-eighths of an inch bit."

I slept sitting up in the visitors' room. I kept remembering the way he had looked when we hoisted him into the ambulance at Deerfield, with the gray blanket across his face.

* * *

Johnny made a very brisk recovery. Six days after the operation he ate a beefsteak sitting up; on the eighth day he was busy drawing a series of parabolas and on the eleventh day he walked the length of the corridor alone. But do not think he didn't suffer.

His eyes were stuck absolutely shut the first day or two; and he spent forty-eight hours fearing that he was blind. The very day after the operation he asked me to bring his physics text to the hospital, and then demanded that we read him the questions at the chapter ends. Thank goodness, he knew the answers! He thought, since something drastic had happened to his brain, that he might have lost his memory. Then he announced suddenly that he knew what he had—schizophrenia! He thought he was crazy. This was because he had once read in a medical book that electroencephalograms and similar tests were used to treat schizophrenia. We brought Traeger in right away, who showed him that although he was dead right in associating this test with schizophrenia, he had nothing at all to fear inasmuch as this disease is never treated by actual operative procedure—and certainly he had had an operation! But Johnny's torture left us blanched. On successive days he thought he was going to lose his eyesight, his memory, and his mind.

But nothing whatever had gone wrong with his faculties. One evening, within a week of the operation, he listened to the Quiz Kids on the radio, and was quicker answering one mathematical and one historical question than the children on the program. He read in Bertrand Russell's *ABC of the Atom* and once he asked Frances to make clear to him the distinction between the words unmoral, immoral, nonmoral, and amoral. He worked out equations of a sort far too abstruse for me, and kept asking for more and bigger textbooks.

Later he became depressed and excitable and worried about the schoolwork he was missing. He wanted to demonstrate that he was far beyond the simple mathematics he was taking. We, on the other hand, told him that routine cramming was part of every education and that he might as well face this now as later. Suddenly he announced on May 10—twelve days after as serious an operation as a human being can undergo—that he wanted to write a letter. Frances thought it might be to Mr. Boyden or to one of the Deerfield boys. But it was to Dr. Einstein. Frances had to persuade him to dictate it to her, and she took it down word for word, comma for comma, exactly as he said it; then he was surprised and impressed at the way it came out:

DEAR PROFESSOR EINSTEIN,

For some time during free periods at school, I have been struggling, I am afraid rather unsuccessfully, with Eddington's *Space, Time and Gravitation*, and the rather fantastic idea occurred to me—here comes the presumptuous part—whether it would not be reasonable to assume that the number and curvature of dimensions of the universe be considered, if not variable, at least "relative." The properties of an event would then be determined by the number and curvature of the dimensions which govern it. Electric and magnetic attraction would then be considered merely a type of gravitation through electric and magnetic "dimensions." This attraction would then continue to follow your law of gravitation. Of course this theory would have to be tested by determining whether it checks mathematically with the equations of Clark Maxwell, and your law of gravitation. But unfortunately I do not yet have the mathematical training to compute this problem.

If by some wildly and impossibly fantastic coincidence, this weird idea should coincide with any ideas you may have had or will have, please do not think of giving me any credit for suggesting it, as I would not deserve it any more than would Newton's apple for catching the great scientist's eye.

Thank you for reading this.

> Sincerely yours,
> JOHN GUNTHER, JR.

I sent this letter to Einstein with a brief covering note and, making us very happy, Einstein replied to Johnny promptly as follows:

DEAR MR. GUNTHER:

I have read with interest the remarks contained in your letter of May 10th. They were, however, too brief to give me a clear understanding of your ideas. I hope to see you when you have recovered so that we may have a conversation about it.

With my best wishes for your speedy recovery, I am,

> Yours very sincerely,
> ALBERT EINSTEIN

When we gave way to the temptation to boast about this correspondence, Johnny begged us not to. His subsequent acknowledgment was, we thought, a model of grace and gentlemanliness:

DEAR PROFESSOR EINSTEIN,

Thank you so much for your very kind letter. I am afraid that in my letter I must have implied that I know a great deal more about this whole subject than I actually do.

I left the hospital over a month ago, and have been trying to gain some understanding of some of the technical terms which I used so freely in my letter.

I shall continue to study, and one day I hope to have the great honor and privilege of meeting you and having the conversation which you so graciously suggested.

We did not know whether Johnny's original letter made sense or not, and so I sent a copy to my friend Francis Bitter, Professor of Physics at the Massachusetts Institute of Technology. He answered:

As for Johnny's letter, I am not certain that I quite understand what he has in mind—particularly as to the number of dimensions of space being variable. The fundamental problem which he proposes, however, namely the unification of equations governing electromagnetic and gravitational fields, is one of the standard headaches of physics, and goes by the name of the Unified Field Theory. Einstein, as well as many others, have worked at it without achieving completely satisfying results. I have never taken up this subject myself. It is amazing that Johnny should be aware even of the existence of this problem.

Johnny was shy; he didn't dictate easily. It took a lot of hard coaxing by Frances to get out of him the letters he did write, some of which appear later in these pages.

Of course Johnny did not know the full seriousness of his illness. Above all we had to shield him from definite, explicit knowledge, since his greatest asset by far—his only asset aside from his youth —was his will to live. But there are layers in a sensitive mind; deep down, he had a pretty complete awareness of what was going on. Very seldom, however, would he allow that lower layer to express itself. Once (neither Frances nor I was in the room and the nurse told us later) a great spray of flowers arrived from Cass Canfield. Johnny said abruptly, "Why don't they wait to send flowers till I'm dead?" But a few minutes later when I came into the room he brushed this mood off with, "Pop, when are you going to get that new advance out of Cass?"

Immediately Putnam broke to him what he had, Johnny did two things. First he called an adult friend, Lewis Gannett, and told him almost proudly, "They drilled three holes right through my head." Second, he telephoned a schoolmate, Edgar Brenner, "Guess what! That pressure I had was a brain *tumor!*"

He announced one afternoon when I came in, croaking, "Wonderful news—I've had another spinal tap!" Almost always when I called him early in the morning to ask how he was feeling, he would answer, no matter how feeble his voice was, "Simply marvelous!" When he learned that what he had was more serious even than polio, he was impressed and pleased. He would laugh and say, "Nothing can hurt *my* old brains!" He was sickish one afternoon and I told him that the medicine he was taking always knocked hell out of people. "It won't knock hell out of me!"

"That child," one of his nurses told us, "is a picnic twenty-four hours a day." When his eyes were still black, he asked for a couple of beefsteaks to put on them, and then suddenly reconsidered. "Oh, no, there's a meat shortage still!" If no one answered his bell, he would alert the entire cor-

ridor by blasting away on his recorder. One nurse
said something mildly anti-Semitic once and he
turned to her gently, "I think you ought to know
that I'm half Jewish." He wasn't hungry for breakfast one morning, and the nurse, by some maneu-
ver, got him to eat some bacon he didn't want. He
turned to her with the words, "You are truly
Machiavellian, like the British in India." He
always wanted to know, with fierce insistence, the
exact nature of his medicaments; he would worm
the information out of the nurses, and then get
more information by a simple process of blackmail
—he would threaten to tell the doctors "on" the
nurses. Once—supreme moment—he turned to
Putnam himself and threatened, if Putnam didn't
agree to something he wanted, to go to an osteo-
path.

And his considerateness! Once he told Frances
that he had been chilly during the night. She asked
him why he hadn't called the nurse. His reply was
that he hated to disturb her because she must be
tired out. One night he telephoned me very late,
when I was asleep. He apologized, "Oh, I'm *so*
sorry, Father!" and then talked a bit in what
Frances once described as "his sweet, gay, buoyant,
breaking voice."

A child knows everything, and nothing. Johnny could tell me all about the Andromeda nebulae—but he didn't know that the simple reason why his head had to be elevated was to relieve pressure. Some days after another spinal tap, Traeger explained the procedure and how the fluid had to be drawn out with extreme care and very slowly; otherwise the brain might, as it were, crash down through its own cushion of fluids. "Well," Johnny said, "Now for the first time I'm glad a doctor didn't tell me *every*thing; I would have been somewhat alarmed if I had known that when the first tap was taken!"

Slowly, very slowly, Putnam took us into his confidence. We were crazily optimistic for a week or two; we even made plans for Johnny to recuperate down at Virginia Hot Springs. The preliminary diagnosis was of astrocytoma. This is a type of tumor comparatively benign, and even if Putnam didn't get it all, there was a fair chance that X-ray therapy might knock out the rest. It takes some weeks to prepare the slides of tissue that confirm the type of tumor; the procedure is extraordinarily involved. Not only do tumors grow; they may change from one type to another. A doctor told me in the corridor one day that the slides

[53]

were ready; I saw something in his eyes and could not bear to ask what they indicated. I walked right by him refusing to talk and he must have thought I was crazy. Then the next day Putnam called me into his office, with Lester Mount, his youthful associate who carried on to the end, standing by. What Johnny had was much worse than they had feared; it was not an astrocytoma but something called an astroblastoma undergoing transformation. Now Putnam explained why he had not closed the skull but had left open, covered only by a flap of scalp, an area in the head about as big as my hand. This was to provide "decompression," i.e., to allow the tumor, if it continued to grow, to bulge outward, instead of inward which would destroy the brain. If Putnam had sealed up the skull with bone or a plate, Johnny would have been dead in a month. The scalp is quite a tough piece of leather but, obviously, it provides less resistance to a growing tumor than would bone. The flap—what agonies Johnny had over that flap for fifteen months!—was a kind of escape valve. Also to have the skull open (except for this flap of scalp) allowed easier access to X-ray. But the open skull produced disadvantages too, in that

Johnny had to be very careful not to fall or otherwise injure the soft spot. The soft spot was shockingly large; when the bandages were removed and we saw for the first time the extent of Putnam's incision, we were stupefied.

I went to the library to do some more reading. There I discovered that any type of tumor bearing the prefix "glio" (=glue) is invariably fatal. I rushed to several doctors, appealing to them to say that surely, surely, Johnny's tumor could not be of *this* deadly type. "Of course it isn't," they replied.

Johnny continued to recover nicely. His eye didn't droop any more and, except for a slight weakness in his left leg, he seemed to be quite well. The weakness bothered and worried him, of course, but Traeger explained it as purely the result of shock—surgical "insult" and consequent edema. Over and over we told Johnny, lying, that the tumor was dead, that Putnam *had* got it all. What was affecting him, we said, was simply the after-effect of a terrific operation. Once he seemed sad. He muttered bleakly, "Everything is frustration!" Then he snapped out with the remark, "I've done nothing for sixteen years except think about myself, so naturally I sometimes get depressed!" Always he tried to save *us* from worry.

What really interested him was getting back to school. He was terrified to think of what class work he had missed, and how he was going to manage to catch up. School!—we did not dare tell him that it would be a miracle if he ever saw a school again. How he fooled us on this!—as these pages will soon tell. Hoping with such vehemence to recover, yearning with such desperation to be all right again, refusing stalwartly to admit that his left hand, too, was showing a little weakness now, he became heartbreakingly dutiful about everything the doctors asked. He was still limited as to fluids; drop by drop, he would measure the exact amount of water he was permitted. All he wanted was to obey, to obey, and so get back to school.

Once, however, he had an outburst. In a frightening and intense moment, with his blue eyes glowing and burning under the white turban of bandage like Savonarola's, he protested with the utmost violence against his regime at Deerfield. He said that his schoolwork "went against the grain," that he couldn't stand being held back by his class any more, that he must get on to Harvard right away because he was being "held back," that he was perfectly competent to do college physics now but that he had to waste all his time and energy on "stultefying" texts and experiments that he had long since passed by. "I feel a moral conviction about this," he exclaimed to Frances, "a religious *conviction!*"

He began to show great curiosity now about what caused the tumor, and he even suggested that the strain of "holding himself back" at Deerfield could have caused it. What did cause it? Patiently Dr. Mount and the others traced back through Johnny's whole history for any evidence of a shock, blow, or other clue. Once Johnny said triumphantly, "I know what caused it!" "If you do, you'll have revolutionized medicine," Mount replied with his grave, friendly voice. Johnny's theory was that he had been sitting far back in a chair playing chess and then slipped and banged his head on the iron radiator. But this blow had not even left a bump or bruise, and nothing so slight could possibly have put into motion any growth so deadly. The plain fact is, of course, that nobody knows what causes a malignant tumor. The origin of life itself is not more mysterious. The causation of cancer is the greatest and most formidable of all the unknowns of modern science.

One grayish afternoon Johnny showed this prayer to Frances:

> Almighty God
> forgive me for my agnosticism;
> For I shall try to keep it gentle, **not cynical,**
> nor a bad influence.
>
> And O!
> if thou art truly in the heavens,
> accept my gratitude
> for all Thy gifts
> and I shall try
> to fight the good fight. Amen.

The story behind the prayer is this. He called it an "Unbeliever's Prayer." Johnny had never prayed; perhaps this was a reaction from his dislike of chapel at Riverdale and his resentment at having been obliged to spend a good deal of time listening to organized religious exhortation. To counteract this tendency, Frances began to read him prayers of various kinds—Hindu, Chinese, and so on, as well as Jewish and Christian. He was interested in all this, but it did not mean very

[58]

much to him at first. Then she started him on
Aldous Huxley's anthology of prayer, *The Peren-
nial Philosophy*, and told him how intimate and
very personal prayer could be. Once she suggested
that if it should ever occur to him to think of a
prayer himself, of his own special kind, he should
tell her. So, very casually, with an "Oh, by the
way . . ." expression, he said, "Speaking of prayers,
I did think one up." He recited it and only disclosed later that he had previously written it down
himself and memorized it.

At about this time he became fascinated by the
Book of Job. He asked Frances to read it to him
several times—which she did while barely able to
face doing so. "It will teach me patience," Johnny
said.

He was cheerful again that evening. "Pop," he
said, "you should be working on *Inside U.S.A.*
and writing speeches, not spending fifty thousand
dollars a month to keep me here!"

Frances always told him that when I arrived he
should cheer *me* up. By this time, very sensitive to
us both, he had developed the habit of presenting
us with different sides of his character. Often
Frances and Johnny, talking about philosophical

[59]

abstractions or merely discussing hospital routine, would hear me come along the corridor; Johnny would instantly change his attitude and demeanor, even his physical posture. Similarly, when I was alone with him, he would make an about-face when Frances came. She helped him immeasurably by trying to make theater of the recurrent daily medical crises, and he played up to this wonderfully. He dramatized his relations to the doctors and the nurses. For instance if they seemed tired and overworked, he took pains to be particularly cheerful with them; if not, he would assault them with amiable complaints.

On May 28 we had more bad news; Johnny fainted going to the bathroom, and the pathologist's report was worse. We tried to face asking the questions we could not bear to ask, since little by little new horriblenesses, new dreadfulnesses, were being hinted to us—about blindness, about paralysis. Several doctors seemed to be avoiding us, and Putnam himself said little, glossing it all over, and telling us to put our trust in the X-ray therapy now beginning. One afternoon we walked briefly in the garden, and a hundred yards away in a doorway leading to the lawn, I saw a young man, apparently a spastic case, writhing, twisting, grimacing, being held tight by a male nurse. I turned Johnny around sharply. I do not know whether he saw or not.

Later I peeped at the sheet on the technician's desk in the X-ray room. There it was as clear as daylight—Johnny's tumor was "undergoing glioblastomatous transformation." That prefix "glio"! No doctor had quite dared to tell us.

The first time I saw Johnny really frightened came at about this time, when he got ready for the first X-ray. He kept saying that "surely" this must be "just for taking pictures." He said to me again and again, anxiously, "It's just for *pictures*, isn't it?" Then he knew from the time he spent under the machine that something much more serious than taking pictures was going on, and that this must be a form of treatment. He turned to me firmly and asked, "Does this mean that I have cancer?" Then he murmured to Frances later, "I have so much to do! And there's so little time!"

CHAPTER 2

OHNNY was discharged from Neurological on June 1, and he moved to our apartment. But he had to return to the hospital every morning for X-rays until June 20 when he was cleared to go to the house in Connecticut, a hundred miles away, for what we hoped, even then, would be an uninterrupted quiet summer. The X-rays took a frightful lot out of him. Several times it seemed that he simply could not manage, physically, the brief walk down the corridor to the elevator and the few steps into my car. The radiologist would not allow more X-rays after June 20, though it was too early to tell whether the tumor was diminishing or not. Too much X-ray will kill any tissue, as surely as will a tumor. Also it can have a vicious deteriorative effect on the white blood cells, the bone marrow, and, as everybody knows, the skin. The radiologists had to estimate almost as carefully as Putnam in his operation exactly how far they could go before the beneficial effects of X-ray might be outmatched by their destructive force, both on Johnny's scalp and on the brain itself.

Those were difficult and unhappy mornings—the traffic-choked drive across the city and then up the West Side Highway; slow, careful guiding of Johnny into the building, and the long waits for the slow, inefficient elevator; the technicians helping him up on the table, and then exposing his head to the machine as the switches clicked and that tremendous instantaneous power leapt out invisibly into the skull; then putting the bandage back, which I had learned to do; then talks with the doctor for a while and then the ride home again. "Papa, I feel so *sick!*" Johnny said on one of the rare occasions when he would even admit that he was ill. He had now built up a secret defense within himself about the X-rays. It was that they had nothing whatever to do with that little "wart" (he had decided now that of the original tumor a small "wart" remained) but were merely to ameliorate the general postoperative swelling.

[63]

His good humor was equaled only by his courage. Almost six months later, he confided to me, "You know, Father, I was so worried during those X-rays that I couldn't sleep at night. I almost gave myself ulcers." He did sleep, though.

The first minute he was home after the operation he did what we had anticipated—dived for the Britannica to look up brain tumors. We had taken the precaution to hide this particular volume because, among much else, the article said that almost all brain tumors end with blindness. I cannot recall now how we explained its absence. Johnny fumed for a while and then resigned himself to the mysterious ways of parents. That evening we discussed plans for the country calmly. Indeed, he had talked them out with Putnam himself. Swimming, diving, boating were forbidden—this last a cruel blow—because of the flap, the soft spot, in the skull. Johnny could still have had no real perception of the intense seriousness of his illness, because he had asked Putnam, "But can I go rock climbing?" We were told that he must rest above all, take short walks, read in moderation, and exercise the left fingers and left toes. The weakness here was gradually becoming more pronounced,

[64]

but we did not know, nor did the doctors, whether the tumor was responsible, or, as may have been the case, whether it was a temporary setback caused by the massive shock of the X-rays. Probably it was at this juncture that we first became seriously impressed with what little doctors do know. To question after question—what about the eyes? what about special therapy for the fingers? what should we particularly watch for or guard against? what if there is sudden increase of pressure? how long will it take for the glioblastomatous changes to develop? when will it be safe to give more X-rays? what shall we do next?—the answers, despite the utmost good will, were confused and contradictory, simply because the course of any brain tumor in a child is unpredictable. Of course we expected too much. But it was our worst burden that we were never sure about anything, not merely from one day to another, but from one moment to the next.

That whole summer, the summer of 1946, is a spotty haze in my memory. Mostly we were moving Johnny in and out, because he had to return to New York every ten days or so for checkups and to meet successive crises. But as to our own emotions I am trying not to write about them. What

terrors and horrors of anguish it meant to Frances,
I leave to the imagination.

The house at Madison was, from almost every
point of view, perfect for a convalescence. It immediately faces the Sound, with a broad scallop of
private beach. Johnny's room was upstairs and,
until the end of the summer anyway, he did not
have too much difficulty getting up and down.
Here were his most precious books, his specimens,
his gramophone records, all the paraphernalia for
his studies. Downstairs he could loaf on the long
balcony, putter about in the sand, lie with Frances
in the sun, make barbecues, wade, and even play a
little with his boat. His workshop and laboratory
were in the garage a hundred yards away. Here
were chemicals, mechanical apparatus, rocks, laboratory equipment, and the vast heterogeneous
assortment of things a boy collects and works with.
Johnny was not bedridden. He never was that summer. There was not a single day in which he did
not spend an hour or so in his workshop. He demanded that he be allowed to do his work in his
own special way. Thank goodness he was well
enough for that. He loved every minute of that
summer. Frances was with him all the time and I

came out weekends. Our friends dropped in and
Frances's brother and my sister were steady
visitors. We had a Japanese cook who enchanted
Johnny not only by the splendid things he made
to eat but by his skill arranging flowers from the
garden. "Tom," Johnny announced to him one
evening, "you are an artist." But I do not know
whether it was a lamb chop or a spray of roses that
he was referring to.

Particularly I remember Johnny's considerateness, even when he got sicker. Of course he wanted
his classmates and other friends of his own age to
come for weekends, and several did. But he would
hesitate to ask them, for fear they might be bored
—inasmuch as he himself could not join them in
sports or outdoor games. He was vehemently worried that his illness might upset our future plans
and about how much it was costing and about
Frances's work and my book. I was less than half-
way through *Inside U.S.A.* and hopelessly behind.
Johnny knew that the deadline for delivering the
MS was October 1, and he knew perfectly well that
I could never make it. His first question when I
came out, expressed with lively irony and disbelief
in the veracity of what I would tell him, was, usu-

[67]

ally, "Well, how many whole chapters did you write yesterday?" Then: "You'd better hurry!" As to the work Frances was doing he would remonstrate with her gently and then encourage her, "Remember your destiny, Mother!"

This is from the diary Frances was keeping:

Yesterday, wet, cold. Bought peaches and strawberries. Johnny dressed and cheerful when I arrived. Read *Henry V*—Johnny read aloud the great speech. We recalled English chronology. Nap before dinner. . . . John came at eight. We played "Twenty Questions." John's thought: *Henry V* film; Johnny guessed it. Mine: top button on John's pajama jacket. Johnny's: Prof. Einstein's signature; we didn't guess it.

These were some items from his conversation at about this time, as taken down by Frances:

I don't know what I'd do without you, Mother.

I think I'd like a bottle of champagne at school for my birthday.

My thinking is independent of my temperature—it just depends on my stimulants.

I'm going to write a theme "On Being a Guinea Pig," with teleological aspects.

I handed in my theme on theories of homework just

two days late—as proof of my argument and justification for not having done it on time.

In my Fourth Year, I'd like to take just Math, with a tutor, and Relativity with Professor Einstein.

A major problem continued to be what to tell him. If the tumor was indeed mostly gone, how then explain the continued bulging? But beyond this there were larger questions. *Why* was Johnny being subjected to this merciless experience? I tried to explain that suffering is an inevitable part of most lives, that none of this ordeal was without some purpose, that pain is a constituent of all the processes of growth, that perhaps the entire harrowing episode would make his brain even finer, subtler, and more sensitive than it was. He did not appear to be convinced. Then there was a question I asked myself incessantly. Why—of all things— should Johnny be afflicted in that part of him which was his best, the brain? What philosophical explanation could one find for that? Was all this a dismal accident, purely barren and fortuitous? Beethoven was struck deaf and Milton blind and I met a singer once who got a cancer of the vocal cords. But if the connection of circumstances was not fortuitous, not accidental, where was justice?

Johnny said to me once, "The worst thing is to worry too little, not too much. Let's keep up a tension." It was as if he were girding himself for the struggle only too obviously under way, between life and Death.

Crisis followed crisis now in a series of savage ups and downs. The flap, which we called the Bulge or the Bump, got slowly, mercilessly bigger, until it was almost the size of a tennis ball sticking out of his head. "Oh—it's the way those things go after X-ray," we tried to explain it away. He accepted this—perhaps—and was cheerful and determined. Meantime we came to learn a new medical word that pursued and haunted us for almost a year—papilledema. This means, to put it roughly, a forward protrusion of the optic nerve, which is an extension of the brain itself. When pressure exists inside the skull, causing damage to the optic nerve, the amount of injury may be calculated with an ophthalmoscope. When the eyes are normal, the papilledema is zero. Then degrees of injury are measured in an ascending scale to 10 diopters. The higher the papilledema, the worse the situation. Before the operation Johnny's papilledema was a full 10; when he left the hospital, it had dropped to 2. Now it hovered between 4, 3½, and 4 again. Always the first thing a doctor did was to measure this wretched papilledema. Another frightening factor was that though most of the cranial nerves were still normal, there had come a slight lag on the left side of the mouth. Also Johnny had lost a shocking amount of vision. His eyes were more or less all right when he looked straight ahead, but what are known as the visual fields had become sharply restricted, and he could not see well to the side. The doctors called it "a left homonymous hemianopsia." It was as if he had an invisible blinker on the side of each eye.

Nobody should get sick in or near New York in July or August. Putnam and Mount were both on holiday. On July 12 I brought Johnny in to Traeger for a check. Traeger did not like the look of the bump and sent us to Masson, who had taken Putnam's place at Neurological. Masson would not see Johnny that day, which meant that he had to undergo the exhausting business of a hundred-mile drive back to the country, then into town again, and then back once more. Then Masson took one look and said flatly that Johnny could not live more than a couple of months. That ugly analogy came up again—his brain was like an apple with a spot in it.

On Wednesday, July 17, I was back in town. At about midnight Frances called from Madison. The bump had opened and was leaking pus.

From that day until a month before he died, there was never a single day in which Johnny— patient, brave, humorous—did not have to go through the laborious nuisance and ordeal of having his head dressed and bandaged.

This leak in the bump was not in itself serious; it was a "stitch abscess" caused by a couple of tiny stitches left in the original incision. But it worried our local doctor in Madison, and Johnny began to take penicillin to avert infection in case the wound should widen where the bulging scalp was stretched thin and taut, or in case there should be another ulceration. We prayed that this would not happen. It happened, though. The bump burst open in another place. I got another hurry call from Madison at noon on the twenty-fourth, drove out at once, and Johnny was back in Neurological that evening. He was not worried much; only annoyed that his work was being interrupted. But he was skeptical when we said that he would be hospitalized only for a day or two. "Once they get me here," he declared, "they keep me."

We had heard, meantime, about a magician

named Wilder Penfield in Montreal. Half a dozen folk had suggested that we get in touch with this renowned surgeon who, like Putnam, has international rank and who, like most great brain surgeons, is a poet. Traeger tracked him down, and he agreed to come to New York to have a look at Johnny. It was interesting to notice how impressed Neurological was with Penfield. The manner of the whole sixth floor abruptly changed. Previously Johnny had been a hopeless case; now he was a phenomenon of considerable interest. Putnam interrupted his holiday, and he, Penfield, Traeger, and another doctor spent the morning in consultation. We had told Johnny casually how eminent Penfield was, and his greeting to him was quite characteristic. He measured Putnam and Penfield together, and then asked, "Where's Cushing?"

Frances often helped Johnny to time or rehearse his little jokes, but this one caught us unawares. He knew perfectly well where Cushing was. He was meeting two of the three greatest brain surgeons, and would probably be meeting the third quite soon.

Penfield said something about the fine recovery he was making, in order to reassure him.

Johnny replied, "I'm not so much interested in

the spectacular nature of my recovery as in the exact seriousness of my complaint." That held everybody for a while.

Penfield spent an hour on the slides; always, in a thing of this hideous kind, the possibility exists of mistaken diagnosis, and the tumor might have changed for the better or worse. We waited, and then with everybody listening Penfield cut through all the euphemisms and said directly, "Your child has a malignant glioma, and it will kill him."

He wrote in longhand on Johnny's chart:

The neoplasm is obviously a malignant glioma. The removal and decompression has given him some longer lease of life, and it has been a happy interval. The presence of two small cysts within tumor, as proved by Dr. Putnam's puncture, is consistent with glioblastoma and it is apparently present beneath and outside the dura. The scalp defect is obviously the result of pressure necrosis, not primary infection.

I would recommend healing the area if possible in a few weeks. Further X-ray treatment only when radiotherapist decides the skin and brain will not suffer from it.

If operation is decided upon, occipital lobe amputation might be carried out with some sort of skull closure. This would not prolong life much if at all.

It might make him able to be up and active over a greater portion of the life.

I can see nothing that could have been done up to date that has not been done. This is the tragedy of such cases.

Of course no one told us that complete occipital lobe amputation would mean blindness.

We asked people the next few days how the end would come, and once more new horrors, new dreadfulnesses, were disclosed. One of the nurses said that in tumors of this type the patient gradually lost all function, even that of control of his own secretions, and died in the end like a kind of vegetable.

Johnny did not lose function. He lived almost a year after this, and he did not die like a vegetable. He died like a man, with perfect dignity.

*　　*　　*

Now we struck out hard on new paths. The rest of the summer is the story of pillars in a search. There might be some ray of hope somewhere despite Penfield's death sentence. But we must act quickly. Frances thought that physicists or atomic scientists who worked in the medical field during the war might have discovered something new about brain tumors unknown to the public at large, and I wrote or telephoned to doctors all over the country to investigate this possibility. The thought never left us that if only we could defer somehow what everybody said was inevitable, if only we could stave off Death for a few weeks or months, something totally new might turn up. What we sought above all was time. Our search was, to put it mildly, further stimulated because at least two doctors, after the Penfield consultation, urged us to put a cap in Johnny's skull, which would eliminate the bump. Also, by driving the tumor inward, it would kill him. Euthanasia is, of course, forbidden in the United States. But the doctors wanted to be merciful.

I wrote to Hutchins at the University of Chicago, to both Lawrences at the University of California, to the head of the tumor clinic at Michael Reese in Chicago, to a splendid physician in Boston who had just come back from Russia, to one specialist who was experimenting with radioactive phosphorus, to the head of Massachusetts General, and to our friend Professor Francis Bitter. We asked one and all the same question—did they know

[76]

anything new? Was there any hope?—particularly in developments in medical physics. One and all made the same reply, in painstaking and courteous terms, that nothing at all was new, that Johnny was having the best and most expert medical care the entire world of science could provide, that no new discoveries at all had come in this field, and that, therefore, hope was nil.

One morning Frances found an item in the Sunday *Times*, hardly two inches long, describing some remarkable ameliorations of tumors—not brain tumors, but just tumors—caused by intravenous dosages of mustard gas.

This is, of course, a deadly poison. Scientists had come across it as a possible treatment of cancer directly out of military experiments. Mustard gas kills by attacking certain cells with abnormally fast growth. What is a tumor if not something in the body growing fast? Hence the transposition was easy to the hypothesis that mustard, or HN_2 as the doctors called it, might conceivably pick out and attack tumor cells, while not harming appreciably other cells, if administered in tiny doses with great care. Moreover the researchers had discovered that mustard had mysterious and extraordinary effects on various other elements in the body. It seemed just the sort of thing we had hoped the scientists would tell us about. None of my eminent correspondents had so much as mentioned it. But there it was plain as day in the New York *Times*.

Frances, through friends in New Haven, set out on the trail of this mustard. We chased it to the University of Utah, to an experimental station in Maine, and to the offices of the American Cancer Society. After a week we tracked it down finally at Memorial Hospital, New York City—ten minutes' walk from our apartment. What decided us to use it was the word over the telephone of one of the most celebrated physicians in the United States: "If it were my son, I'd try it." And certainly there was nothing to lose. Nothing at all to lose!

Traeger got in touch with Rhoads, the head of Memorial, and I went to see Craver, the medical director there, who put at our disposal Dr. Joseph Burchenal, a young scientist with a fine war record who was in charge of the HN experiments. He drove out to Neurological with me, and we put it up to Mount. Now it is a ticklish business to mix up hospitals. It is a very rare thing for a doctor affixed to one hospital, like Burchenal at Memor-

ial, to do work at another like Medical Center.
Let me thank everybody who generously helped
waive the rules. Within twenty-four hours of first
talking to Craver at Memorial, I saw the first injection of mustard gas ever given at Medical Center administered to Johnny. It was all so im-
promptu and urgent that I myself carried the
precious, frightfully poisonous stuff from one hospital to the other.

* * *

During all of this Johnny was reasonably confident. At I do not know what cost to his inner
resources, he maintained the boldest kind of front.
Once Bill Shirer and the late John T. Whitaker
dropped in; each had just had a serious hospital
experience. "What did you talk about?" I asked
Johnny when they had gone. Reply: "It was very
boresome. We discussed our operations."

Frances gave him some science fiction once. "The
trouble with science fiction," Johnny said, "is that
it's bad fiction and no science." He announced one
morning that he wanted to be five things—a physicist, a chemist, a mathematician, a poet, and a
cook. He added soberly, "And since I'm only six-

[79]

teen, I think I have a good start in all." Once he asked for a bath after dinner, took it, and later congratulated Frances on her self-restraint in not coming in to wash him!

She arrived at the hospital as usual at noon one day, and he wasn't in his room. She rushed down the corridors and found him out in the garden, all dressed, lively and triumphant. "I escaped!" he told her with great satisfaction, as a worried nurse came up. It was at about this time, too, that, discussing his various doctors, he said, "Maybe I will be a historic case!"

But after Penfield's visit he was very wan and dispirited. He would stand in the doorway and look at us tentatively, appealingly. When he telephoned in the mornings and evenings, his voice had no body. The frightful strain had begun to drag him down. Half a dozen times, when Frances tried to keep him from doing too much, he would exclaim again in protest, "But, Mother, I *have* to get my work done!"

He could not have survived this summer had it not been for his mother's brave and understanding spirit. So that he would not be frightened she talked to him as if casually about the narrow es-

capes she and other people had had from Death,
and it relieved him greatly to learn that several of
those whom he loved had *almost* died. She made
the most of every medical ritual, and taught him
to squeeze out of every conceivable occasion, no
matter how painful, every atom of humor possible.
She read him poetry on meditative and religious
themes, and he made his own anthology of poems
he liked by reciting them into a transcribing apparatus, and then playing them back when the
mood was on him. Here, too, the sharp demarca-
tion he made between Frances and me, based on
his solicitude for us, became manifest. With Frances he talked of Death often; with me, almost
never.

* * *

Johnny got his first doses of mustard between
August 1 and 5. It had never been tried on a brain
case before. Usually mustard makes a patient very
sick at first. Also there was considerable local pain
in that the veins in his arms were difficult to find,
and the injections produced heavy bruising.
Johnny puked plenty the first day, but not after
that. Then there had to be a close watch on after-

effects, since one of the results of mustard is to drive the white blood count down. The figure may drop alarmingly, enough to scare out of his wits any doctor who does not know what is going on. The white blood corpuscles serve an important function in combating infection, and so it was necessary to keep dosing Johnny with huge amounts of penicillin too, as compensation for the temporarily lost white cells. When we drove up to the country we filled a rubber bag with dry ice and chucked the penicillin in it, and for over a month Johnny had to have a blood count every day, which was still another item in the onerous routine he had to undergo.

"How's my blood, Father?" he would ask.

"Fine."

"Let me know if it goes under a thousand."

The first series of mustard shots did Johnny great benefit. Of this there is no reasonable doubt, I believe. They stepped up his vitality and made him fresher, stronger. As to the second series I am not so sure. For we decided on an additional course of mustard, and Johnny had these further shots late in August, when the first results seemed good and X-rays were still precluded by the state of the scalp.

Johnny checked out of this visit to Neurological presently and he was well enough, that same afternoon, to see the movie *Henry V* with Frances and my sister Jean and to walk a few blocks. But there was something sardonic in his last word to his favorite nurse when she said goodbye. "Oh," he waved to her, "I'll be back."

I have before me now a slip of paper on which, that evening, he scribbled down an agenda list for the country; it gives some measure of his ardent hopes and fears:

1	Bandage	Hair
2	Fluids	Horse
3	Sailing	Athletics
	Biclying (*sic*)	Pennicillin (*sic*)
4	Swimming	Bone
	Traveling	Glasses
	Rowing	Nap
	Driving	

Out in the country he picked up quickly. One could see him brace himself valiantly and set about making up lost time. He did schoolwork and for relaxation worked out mathematically all the odds possible in poker, among many other things. Once he listed all his doctors; once, fascinated like most children by the mysterious entity of the family, he drew up his family tree with great elaborateness. One morning in New York I got this letter:

Saturday

DEAR PAP,

Here is the list of chemicals:

1 lb. acetone
4 oz. Amonium Chloride
1 oz. Sodium hydride
U-tube (with arms)
2 ft. thick-walled rubber tubing to fit arms U-tube
2 rubber stoppers to fit U-tube
2 ft. glass tubing to fit *inside* rubber tubing.
1-hole rubber stopper into which glass tubing will fit.
Any test-tube into which stopper will fit

love,

JOHNNY

I scurried around to pick up all this and in addition to find a cargo of dry ice he needed. What shame I feel now that I had never taken this request for dry ice seriously enough! He had asked me for it several times, but there seemed to be more important things to worry about, and I had neglected to bring it. Johnny repeated his request —gentle soul!—but never loudly enough to embarrass me. Finally I brought it. This dry ice

[84]

(enough to fill a bucket) was of the utmost importance. With it he was going to perform an experiment he had been working on, in theory, all summer—the liquefaction of ammonia by a quite new process.

One of Johnny's great friends, and a cardinal influence in his life, was his neighbor Mr. Weaver, who taught chemistry at Andover. For summer after summer, this good and generous man had been Johnny's best adult friend. Mr. Weaver came over and helped him when his own weakened hands and failing co-ordination were not quite up to the mechanical tasks involved. Johnny insulated a big can with rock wool and pumped the gas as he made it through another receptacle filled with the dry ice. The experiment worked, praise be. Never before had ammonia been liquefied in this precise way. Johnny had truly invented something. His pride and happiness knew no bounds—though he scoffed modestly at what he had done.

Frances wrote: "A leaf in the solution freezes stiff, then breaks at the blow of a knife with an icy clink. O triumph! His dark blue eyes shone with joy." That evening when he kissed her good night he exclaimed, "It's been another fine day, Mother!"

There were preciously grasped delights that summer. Once Frances found him late at night intently rearranging his rocks in accurate geological classification. Once she had a party, with the ladies in long dresses and their hair up, and Johnny helped serve the food like a serious, conscientious host. He read Christopher Morley's "On Unanswering Letters" with delight, and one evening I read aloud a Ring Lardner story about a caddie and he laughed till the tears came. Once I gave him a ten-dollar bill and he asked Frances, "Where shall I hide it?" She replied, "In the only place possible—in bed." Johnny: "What a woman!"

Once we celebrated a release from hospital by giving him a spoonful of champagne. He had another spoonful and then announced, "Let's have a little more formality around here. I'm going to call you Mother and Father from now on, instead of Mutti and Papa."

Once Mr. Boyden and Mr. Hayden, one of his favorite teachers, drove down from Deerfield to have a day with him. This was a red-letter occasion indeed, and Johnny talked to them soberly about school the next term. Of course it had oc-

curred to him by this time that he *might* be unable
to get back to school, but the idea was so unthink-
able that for the most part he suppressed it. Mr.
Boyden's visit was a great turning point in restor-
ing hope. Johnny explained to him how he intended to make up the time he had lost, and then
registered frankly all the complaints he had previously made to us. Later he amplified his views.
"It's not that I'm complaining; I was just giving
an explanation of why I don't get better marks."
Still later: "You know, Mr. Boyden is the most
persuasive man anybody ever met!"

He was busy, impatient and irritated but not
discouraged at not being fully well, and packed
with wit. Once he said that for me to lie on the day-
bed was like being a cigar on a toothpick. I was
hungry once and he said, "Give Father three beefsteaks for an afternoon snack—it will help his vitality." One evening the laundry failed us, and he had
to wear a pair of my pajamas. "So," he sighed with
mock weariness, "the dread day has finally come
when I am you." We continually urged him to
rest, and it was a struggle for Frances to get him to
bed at a reasonable hour. "If I hear any more of
that talk about not staying up till twelve, I'll dis-

inherit myself," he would say. One day he said he
had discovered a real reason for the existence of
relatives and inlaws—"For surgeons to practice on."
He often joked about his illness. "I bet that ole'
mustard has knocked the tumor out!"

But on August 31 there was again a new leak in
the bump, and the white blood count was below
1,000. The papilledema was high again, and he
seemed to be fading fast.

* * *

Meantime we were working on another tack.
Not for a moment had we stopped searching. Early
in the summer Raymond Swing told me astonishing stories about a doctor named Max Gerson who
had achieved remarkable arrestations of cancer and
other illnesses by a therapy based on diet. Gerson
was, and is, a perfectly authentic M.D., but un-
orthodox. He had been attacked by the *Journal of
the American Medical Association* and others of
the massive vested interests in medicine; Swing
himself had been under bitter criticism for a broadcast describing and praising highly Gerson's philosophy and methods of dietary cure. My own first
reaction was skeptical, and Frances was dubious

too. Then I learned that Gerson had long experience actually in brain tumor cases, having been associated for years with a famous German neurosurgeon, Foerster, in a tumor clinic at Breslau before the war. I went to see Gerson. He showed me his records of tumors—even gliomas—apparently cured. But I was still doubtful because it seemed to me inconceivable that anything so serious as a glioma could be cleared up by anything so simple as a diet. He impressed me greatly as a human being, however. This was a man full of idiosyncrasy but also one who knew much, who had suffered much, and who had a sublime faith in his own ideas. Frances and I had a long talk with Traeger. At first he violently opposed the Gerson claims, but then he swung over on the ground that, after all, Johnny was deteriorating very fast and in any case the diet could do no harm. I stayed at Madison one weekend and Frances went into New York, visited Gerson herself, and looked over his nursing home. She was impressed too. We made a sudden decision over the telephone. We had tried orthodoxy, both static and advanced, and so now we would give heterodoxy a chance. If only we could stave Death off a little longer! And—once more—there was absolutely nothing to lose.

[89]

I took Johnny out to dinner in Madison and broke it to him that we would be going into town the next day for new and further treatment. This was a grievous shock. It was the first time that I saw him seriously upset. He struggled to keep from tears. He flung himself away from me and crept up-stairs. Mostly this was because he was midway through preparations for another serious experi-ment. But by the next morning he was buoyant again—so much so that I dreaded more than I could tell what would have to be the next bad news broken to him, that he could not go back to school. "I'm sorry I bawled last night, Father," he said with his wonderful radiant smile. Then, limping slightly, he pushed off to his laboratory for a morning's work, sober and dutiful. School— and the science work he was doing—was prac-tically all he talked about.

Frances came out, and on September 7 we drove swiftly in to Dr. Gerson's nursing home. It was a fiercely hot, noisy afternoon, and Johnny was sick with strain, just plain sick. So a new long chapter in his indomitable struggle began.

CHAPTER 3

THOSE September days were grim at first. Johnny lay there pale and panting with misery. His blood count slipped lower and lower, and great bruises appeared on his arms and chest, caused by breakdown of the capillaries. We had been warned that the blood would go very low, and perhaps we were needlessly alarmed— it might well have come back of itself. But anyway we were worried sick. One doctor told us that the reason he had seemed so casual when Johnny entered the Gerson nursing home was his conviction that he couldn't possibly outlast the week anyway. In particular what is known as the polymorphnuclear count of Johnny's blood (I will not go into the technical details) was staggeringly low —down to 3 percent, and the red cells showed a profound anemia. One specialist told us later that

[91]

he has never known of a recovery with such a blood condition.

Within a week, Johnny was feeling, not worse, but much better! The blood count rose steadily, the bruises were absorbed with extraordinary speed, the wound in the bulge healed, and, miracle of miracles, the bump on the skull was going down!

Traeger had walked down the street with me to meet Gerson. He was deeply pessimistic. He said, "We'll move Johnny to a hospital and try massive transfusions—nothing else can save him." The two doctors retired into the kitchen, and came out after half an hour. Then Traeger looked Johnny over slowly and said, "Never mind about the transfusions. Let's do it Gerson's way for another twenty-four hours."

First, Gerson took Johnny off penicillin. This we thought to be a very grave risk, but, he insisted, penicillin could irritate a tumor. Second, he refused to permit any transfusions or other emergency measures whatsoever. What a terrible chance we thought he was taking! Third, he demanded that for some weeks at least Johnny should have rest, absolute rest, nothing but rest, rest, rest.

The Gerson diet is saltless and fatless, and for

a long time proteins are excluded or held to an extreme minimum. The theory behind this is simple enough. Give nature opportunity, and nature herself will heal. It is the silliest thing in the world to attempt to arrest cancer of the tongue, say, by cutting off the tongue. What the physician should strive for, if he gets a case in time, is to change the metabolism of the body so that the cancer (or other affliction) dies of itself. The whole theory is erected on the basis that the chemistry of the body can be so altered as to eliminate disease. Perhaps this may sound far-fetched. But that diet, any special diet, can markedly influence bodily behavior is, of course, well known. Consider inversely how a milligram or so of a poisonous substance, like potassium cyanide, can almost instantly kill a body. How Gerson decided what foods helped to create new healthy cells, as the diseased cells sloughed off, is not altogether clear to me. At any rate the first principle is to make the diet potassium-rich and sodium-free. Gerson took the line that the body spends an absurdly disproportionate share of its energy getting rid of waste, and that therefore, when the body is ill, it will be much freer to combat illness and build

[93]

healthy cells if the amount of waste is drastically
cut down. Hence, as a patient enters upon the
Gerson diet, not only does he subsist largely on
specially prepared fruit juices and fresh vegetables
that burn down to the minimum of ash, but he has
enema after enema—in the beginning as many as
four or five a day, till the system is totally washed
out and cleansed.

Gerson's sanitarium, operated by his daughter
and Mr. and Mrs. Seeley, was run with the utmost
loving care; I cannot possibly pay tribute enough
to Mrs. Seeley and to Miss Gerson for what they did
for Johnny. Also I saw, month after month, a number of Gerson's cases. One patient, it happened,
was an acquaintance of mine of twenty years' standing whom I altogether trusted. I did not know
whether or not Gerson could cure, or even check,
a malignant glioblastoma. I did learn beyond reasonable doubt that his diet did effect other cures.
Gerson himself, zealot that he is, has never claimed
that his diet will "cure anything," as his enemies
sometimes charge. But some of his results have
been astonishing.

One of our doctors (hostile to Gerson at first)
said one evening that September, "If this thing

works, we can chuck millions of dollars' worth of equipment in the river, and get rid of cancer by cooking carrots in a pot."

The regime was certainly onerous. Johnny said wearily after the first week, "I even tell time by enemas."

This is what Johnny had to eat during the next months. For breakfast, a pint of fruit juice, oatmeal, an apple-carrot mash, and a special soup made of fresh vegetables—parsley root, celery knob, leek, tomatoes. This soup he continued to take at intervals during the day, until he had a quart or a quart and a half. For lunch, heaping portions of cooked vegetables, a salad, fresh fruit, the soup and mash, and a baked potato. For dinner, the same. Later he was permitted pot cheese, skimmed milk, and dry pumpernickel. Nothing canned. Nothing seasoned, smoked, or frozen. Above all nothing salted. No meat, eggs, or fish. No cream, butter, or other fats. No sugar except honey and maple sugar. No candy, sausages, ice cream, pickles, spices, preserved foods, white flour, condiments, cakes, or any of the multitude of small things a child loves. Very little water. All the vegetables had to be cooked with no added water or steam, after being washed, not scraped, and without using pressure cookers or anything with aluminum, and the fruits had to be squeezed in a nonmetallic squeezer. Back to nature!

Do not think this was starvation. Some patients gain rather than lose on the Gerson diet. The meals are enormous in size and, as Mrs. Seeley prepared them, exquisitely composed. Then to compensate for the lack of minerals there are injections of crude liver extract every day, and multitudinous pills. These were assembled in a glass dish every morning, in various colors to denote what minerals and vitamins they contained—thirty or more in all. Johnny took niacin, liver powder, lubile (dried powdered bile), vitamins A and D, iron, dicalcium phosphate with viosterol, and lugol. Iodine—in a precisely calculated amount—is essential to the cure.

Johnny's attitude to all this—he was a youngster with a vigorous, healthy appetite—can readily be imagined. He loathed the diet, but he held onto it with the utmost scrupulous fidelity. Carefully he checked off in his notebook the pills he took each day. Once the reason for a thing was explained to him, he faithfully accepted it. The jokes and protests he made were to let off steam, or provide wry humor to the occasion. One evening I asked him if he wanted something, and he replied instantly, "A dose of bichloride of mercury." Once he said that the husks of vegetable in the Gerson soup were deliberately left there as "abrasives to scour out the stomach" and he announced that he had discovered a cure for tapeworm. "Put the patient on the Gerson diet and the tapeworm will evacuate itself in despair."

All that Johnny was really losing was the taste of food. But the monotony depressed him and he sighed on one occasion, "Really, Mother, this *is* too much to bear!" Once as I was leaving for the evening he called out, "Have a big steak for me and come back and tell me all about it!" Later he was worried that when the diet finally ended there might still be a meat shortage. One night his voice was sick with worry. "Wouldn't just a *little* meat strengthen me and help the bruised nerves in my head heal?"

But he was getting better. This, overwhelmingly, was all that counted. The papilledema dropped sharply, and by the end of September the pupils were almost normal—this was an almost unbelievable demonstration of recovery. Moreover the blood count was up to normal, and, incontestably, the bump was smaller. I left Mrs. Seeley's one evening and walked to the corner and had a cup of coffee, almost insane with sudden hope. I was beside myself with a violent and incredulous joy. Johnny was going to recover after all! I thought of all the surgeons and the specialists and how dumfounded they would be; I thought of all our friends and their sympathy and how the burden of grief would be lifted to a whole small community. Johnny was going to pull through, after all, despite everything, and get well! He was going to beat this evil, lawless thing! He'd show the surgeons how a boy with a real will to live could live!

For these extravagant hopes I had, I must say, a certain medical backing for a time. One of our doctors was very optimistic and all were co-operating well. We had feared that, once we delivered Johnny over to Gerson, the more straight-laced physicians might refuse to see him in this so-called black sheep's den. One doctor, out of the whole lot, was in fact angry and did wash his hands of us —temporarily; he said we had spoiled the "controls," as if Johnny were a rabbit. But Traeger

came over to have a consultation with Gerson every
Saturday for months; Putnam, a great man, had no
hesitation in visiting him and later Gerson several
times visited Neurological (which we had been
told would be utterly impossible), and Lester
Mount, who as time wore on became the closest
of all the doctors to Johnny except Traeger, and
who loved Johnny as Johnny loved him, came to
see him regularly, though a surgeon of Mount's
standing rarely makes calls at all.

Parenthetically I might mention the KR episode.
We were committed to Gerson for a time at least,
and he was a dictator who permitted no opposition;
but this did not preclude our fishing quietly in
other waters. Frances in particular never for an
instant gave up searching. She saw in the papers
that Russian scientists, working with a trypano-
some, had successfully treated certain exposed
cancers such as those on breast or nostrils. This led
us into a telephone chase almost like that on the
mustard. I talked to the United States Public
Health Service and to a doctor in Philadelphia
who had actually worked with KR in Moscow. But
there was none available in this country, and it
could not be shipped because it would not keep.

[99]

Again parenthetically, we explored several other long-shot chances, but from these and other forays, nothing at all resulted. As of the autumn of 1946, it was the Gerson regime or nothing.

Our routine was now established. Mrs. Seeley's nursing home was only a few minutes' walk from my flat. Johnny telephoned us the first thing in the morning, and again just before lights-out at night. He was frightfully depressed sometimes, and he hungered for contact with us the first and last minutes of the day. But he would always say, even if his voice was quavering, "I feel *wonderful!*" As a rule I dropped in to see him at noon, and again briefly in the late afternoon or evening; Frances came at lunch and stayed with him until he went to sleep. Soon, we thought, we would move him home—though this would entail a difficult lot of routine because all his food would have to be cooked specially with special paraphernalia. Frances worked on this and got the apartment ready. She moved in presently, and I took quarters in a small hotel nearby. But as it worked out it was several months before Johnny was able to come home.

I read him the typescript of the Montana chapter of *Inside U.S.A.* "Father," he grinned, "it'll sell a million copies." That week too, I took him for a brief ride. He said, "Ah! I feel alive again!"

I asked him what he wanted most to eat when the diet should be over.

"A glass of full milk, an artichoke with hollandaise sauce, spaghetti and meat balls, and a chocolate ice-cream soda."

One night he had a heavy nosebleed, which could have had the most disastrous consequences. Both Gerson and Traeger raced over in the middle of the night, and since Johnny had a slight bronchial cough that might have started the bleeding again, Gerson relaxed his inflexible rule against drugs and let me rush out to a pharmacy open at that hour and bring back some codeine. Johnny was frighteningly tired. He started to cough again at about five in the morning. Mrs. Seeley crept into his room, and he whispered to her, "I'm afraid I'm being too much trouble." She replied with a cheery "Don't worry about me!" whereupon he considered for a moment and then said, "Somebody's got to worry over you."

Once Frances had an argument with the doctor. Johnny told the doctor later, "The way to handle my mother is to stand up to her." Once, smiling, he exclaimed to me as she came into the room. "Mother looks just like a schoolgirl!"

Frances was lifting his morale all the time. She bought bright-colored scarves which he wound around his bandage, and this amused him no end. She read him biographies of great scientists and all manner of news items and stories about science. Another thing that helped his morale was, if I may say so, my book. He was passionately interested in it—almost as interested as I. He knew what a race I was running, against time and circumstance, and constantly he egged me on. He read every chapter in manuscript, calculated its length and weight and how it would fit into the book as a whole, criticized its content adequately, and continually watched out for errors in judgment and inaccuracies.

Miss Gerson's little girl, aged about six, was fascinated by Johnny, and often came in to see him. He was polite, but bored. Girls of six were really not his dish. Once the little girl tiptoed in and asked if it were all right to stay. Johnny replied, "Okay, if you don't compromise me. Keep the door ajar."

Again from Frances's diary:

Today Johnny said, "Oh, Mother—I've been waiting for you—I have a confession to make—You were
right—as usual—about the dancing. At the end of
this term, you know, they have the senior prom. When
I get back to school, what'll I do? . . . I'll *have* to dance!
Oh, Mother, I've been so depressed—" He was cheerful, confessing. . . . Now he said would I practice
with him and I said I'd love to. I was surprised. But
then injections interrupted and dinner. But after
dinner he asked again, and I pushed back the chairs
and rug. And we danced!

Max Kopf came in one day with his cheerful
"Hello, Chonnie" and played him a game of chess.
He played well for an hour, and then got tired;
Max eased off, to try to let him win. Then I discovered that Johnny's field of vision was so limited
by this time that he could no longer see the entire
board at one glance, and hence had to memorize
during the whole game the position in the rook
files. Other friends dropped in during the placid
afternoons, became duly appalled by the number
of pills he had to take, and were cheered up by
him. Clare Luce organized prayers for Johnny—
if that is the proper way to express it—and she
offered us her Connecticut house for his convalescence. On November 4, his seventeenth birthday,

we had a party, and Mrs. Seeley made a kind of imitation ice cream. Johnny adored parties, and that he should have been well enough to receive guests on this his own special day was a happy event. We tried to record everybody's laughter on a soundscriber—the instrument into which, on other days, he sometimes poured his secret fears.

Of course he was fretful sometimes. Sometimes he was subconsciously hostile to me as if out of resentment at my good health. Once he did a modest amount of complaining and then changed mood and exclaimed, "Father, you ought to slap me down when I talk like that!" Sometimes he reminisced about mild escapades at school, and he gave me to understand that he had cut quite a swathe smoking forbidden cigarettes and so on. Once he inquired, "How many cigarettes do you think I actually did smoke, Father?" I replied something or other. Johnny went on: "None."

He never lost his brightness. Frances ordered some handkerchiefs for Christmas as a gift from him to me with the word "Father" embroidered in his handwriting. He asked her how much they would be and she replied, "Nineteen dollars a half dozen for cotton, thirty-six dollars for linen."

Johnny thought a moment and replied: "Let's give him Kleenex."

His views about the tumor had now solidified; he erected a protective rationale which told him (a) the tumor had never been bigger than "a plum"; (b) Putnam had got it out all right; (c) what was happening now was merely to clear up "odds and ends." Nothing could shake him from this belief. "I don't get just *any* tumor," he told Traeger, "it's like me to get a *special* tumor." He would joke about it often enough. He said, "I'm afraid the tumor has gone to my head," and once he chortled ironically to Frances, "I must take my mind off my brain!"

Stoutly he refused to concede defeat. But he was getting more easily tired and he half-dozed a good deal. Once after he tried to do some card tricks and couldn't hold the cards—they cascaded to the floor as he stood up crookedly—he said in a measured, quiet voice, "I daresay my left hand will always be a little clumsy." It was very rare for him to make such an admission. He fiercely picked up the exercise balls which were supposed to strengthen the left fingers, and which never did.

*   *   *

Cancer is a rebellion—a gangster outbreak of misplaced cells. Extremely little differentiates a normal cell from a cancer cell. One theory of the origin of cancer, which I believe Gerson subscribed to in part (of course I am expressing this in very unscientific language), is that during pregnancy a cell may be displaced in the embryo and may lie dormant for many years in the wrong place. Then it may suddenly get kicked loose, so to speak, and break out with savage violence—a cell that is tooth may wake up in the liver, or one that is bone wake up in the lung. The cell, its energies thus fiercely released by some unknown eruptive force, tries insanely to catch up. But it is in a foreign environment and hence destroys relentlessly what it is surrounded by. Gerson apparently thought most cancers have this embryonic origin; hence he excluded from his diet all those items which, by restless experiments with sodium and iodine, he thought might be the factors that *encourage* the fast growth of an embryo. Johnny had, as it were, a fetus growing in his brain.

Tuberculosis used to be called "consumption" because it consumes. It dissolves a lung or bone. But cancer produces. It is a monster of produc-

tivity, creating a voracious growth that eats up the
surrounding healthy tissue. Gerson's first great
cures, which made a considerable impression in
Europe before the war, were of tuberculosis of the
skin. But Johnny's illness was of a type much more
dangerous and implacable.

All that goes into a brain—the goodness, the wit,
the sum total of enchantment in a personality, the
very will, indeed the ego itself—being killed inexorably, remorselessly, by an evil growth! Everything that makes a human being what he is, the
inordinately subtle and exquisite combination of
memory, desire, impulse, reflective capacity, power
of association, even consciousness—to say nothing
of sight and hearing, muscular movement and
voice and something so taken for granted as the
ability to chew—is encased delicately in the skull,
working there within the membranes by processes
so marvelously interlocked as to be beyond belief.
All this—volition, imagination, the ability to have
even the simplest emotion, anticipation, understanding—is held poised and balanced in the normal brain, with silent, exquisite efficiency. And
all this was what was being destroyed. It was, we
felt, as if reason itself were being ravaged away by

unreason, as if the pattern of Johnny's illness were symbolic of so much of the conflict and torture of the external world. A primitive to-the-death struggle of reason against violence, reason against disruption, reason against brute unthinking force —this was what went on in Johnny's head. What he was fighting against was the ruthless assault of chaos. What he was fighting for was, as it were, the life of the human mind.

* * *

By this time a perplexing concrete situation was building up. On one level, the general level, he continued to improve. He was stronger and got dressed in the morning and went out for occasional brief walks. On a specific level, that of the status of the bump, he was not improving. The bump looked awful now, and had become two bumps. Traeger and Mount came to see him together. Traeger thought that he might get some relief if the protuberance were drained; Mount was fearful of doing this because, under no matter what scrupulous conditions, danger always exists of carrying an infection inward. But he agreed to try. Gerson opposed the idea, and we had a long and painful argument. Gerson felt that any anesthetic, no matter how mild, was likely to kill a patient on his diet, since the bodily system had become so purified that narcotics acted as swift poisons. But he agreed finally to permit a modicum of novacaine and Mount made the puncture. The bump had been fluctuant, which was a good sign, but today it was harder than a rock, and he was able to extract only a few drops of fluid. Glumly, we tried to conceal from Johnny the way our faces looked.

The next time Mount came he made an exhaustive test, and I watched it from beginning to end. What the neurologist seeks to find, if possible, is where the damage to the brain is leading. One test always fascinated Johnny. Mount would ask him to close his eyes, then he would put a coin or a safety pin or a button or something similar in his left palm and ask him to identify it. By working the law of averages, Johnny tried to guess what the object would be, when he could not tell by feeling it, and thus outsmart the doctor. Always, too, the doctor tested his grip, examined all his reflexes, made him try to touch his nose with his fingers with his eyes closed, and asked him a series of standard questions—any tremors, any

[109]

nausea, any chills? So far, thank goodness, the replies were always negative. What really bothered Johnny at this time was something that seems almost ridiculously minor—the way his head itched. The wound had to be dressed and treated and bandaged each day—Gerson, with his superb hands, did this with speed and immaculate precision; and of course the hair had grown out, but since regular combing was difficult, a good deal of matter had accumulated in the hair and was difficult to remove. What a blessed day it was when, with great shouts of glee, Johnny was allowed a real shampoo!

It may seem an odd thing to write, but by this time we knew something of how lucky Johnny was, granted that he had to have a brain tumor at all. The tumor was, God knows, murderous enough; but if it had been in another place in the brain it could have produced even more terrible disasters. Some people with brain injuries twitch incessantly; some cannot walk or talk; some can only pronounce *parts* of each successive word; many lose their memory. Above all, Johnny was lucky in that he had no serious pain and was right-handed. Little by little, his left hand was giving way; if it had been the right, his tortures would have been increased manifold. Johnny's injury was on the right side of the brain, and therefore it was only the left side of the body that felt symptoms.

Mount came again, during a period when everything was very good except the bump. Gerson said, "Your son is saved!" Mount said the exact opposite.

* * *

Now occurred the most remarkable of all remarkable things in the story of this struggle. Johnny accepted with disappointment but good spirit that he could not return to Deerfield—I broke the news to him—and he set out diligently to make up his lost school hours by tutoring. He could hardly walk without swaying; he could scarcely move his left fingers; he had lost half the sight of each eye; he was dazed with the poison from the bump; a portion of his brain had been eaten away; and yet he worked.

Frances found him two tutors and set them into their routine smoothly, while Johnny himself planned his daily endeavor like a general directing a battle. He helped map out the lessons himself,

[111]

and knew with complete assurance and precision just what he wanted each tutor to cover in every session.

Here is a letter he wrote to the headmaster of Deerfield, as dictated to his mother:

November 4, 1946

Dear Mr. Boyden,

I am desperately worried about my work. Of last year's physics experiments I did exactly five out of thirty experiments due. Already I am far behind in chem. lab. I've forgotten every bit of French I ever knew, and there is this year's work, too. It is so *hard* to work when one's sick. My doctors, all twenty-three of them, agree that my tumor may have been growing six months or longer. Even in spring vacation my father noticed that I was very tired. I think that most of the year I was very tired, and I had a reputation for being "in a daze."

I am desperate and utterly miserable, since I don't see how I can ever catch up. My English marks were so terrible last year, too. It seems I'm going to stay in this hellish place forever. Its my birthday today (I'm seventeen). I'm going to Harvard next year (I hope) and there will be college boards—and final exams, this year's and last. It's absolutely a hopeless situation. I must implore you to persuade Mr. Haynes to forget about my physics lab, and Mrs. Boyden to excuse me of the chem. lab. that the other boys will have completed.

My English and French notes are lost, and I am desperate.

Please forgive me for unburdening myself in this way, and give my best to all the boys.

Yours very sincerely,
JOHN GUNTHER, JR.

P.S. You may tell Mr. Haynes that I did finally write up two experiments.

P.P.S. All this has been giving me acute and chronic neurasthenia.

Mr. Boyden, great gentleman that he is, not only replied cordially, but sent Mr. Haynes all the way into New York to give Johnny direct encouragement. Mr. Haynes had, I imagine, intended to stay only a moment or two, but he sat with Johnny hour after hour, and the visit helped his morale immensely.

Then Johnny wrote:

DEAR MR. BILL:

I would like very much to make an initial attempt at passing last year's final algebra exam. During the last month or two of vacation I was tutored in intermediate algebra by Mr. Elbert Weaver of Andover. We covered the whole year's work, especially what I was weak in—logarithms and trig. I am afraid that unless I take the exam very soon, I will for the second time forget all the algebra I ever knew.

Therefore, I wish you would send an adequate examination to my father, so he may give it to me, under regular exam. conditions, in the time limits you may set.

I understand there will be no analytical geometry on the exam, but there will be the binomial theorem and factoring of cubes, all of which I have covered. We will send it back to you for correction, and I hope, if I don't do well on it, I may have another chance when I get back to school.

Please give my best to Mr. Boyden.

Then came a second letter to Mr. Boyden:

Nov. 20, 1946

DEAR MR. BOYDEN,

Thank you so much for your kind and considerate letter. My "neurasthenia" is gone and I am afraid that my letter was one of those that should have been written but never actually sent. I had an exceedingly nice talk with Mr. Haynes, and I am sure that I will be able to make up all the necessary work.

I am used to the diet now, and it really isn't at all bad. Every week I feel better and stronger.

I am trying to keep up nearly with my class in this year's history and English while tackling last year's exams one at a time, starting with algebra. Mrs.

Boyden's subjects will be a pleasure any time I do them.

Came the great day when, under honor conditions, Johnny took a preliminary exam to see how well he would do. This was as much of a landmark as any in the whole course of the illness. One of his tutors thought he ought to wait, but there was no holding him. Johnny exclaimed, "Oh, Mother, tell him I want to get that test off my mind—and do the other things—my chemistry and physics— please *tell* him!" He repeated what had become a recurrent remark, with a strange, faraway look in his eyes (a "beyond" look, Frances called it), "You don't understand, Mother; I have so much to do, and there's so little *time!*" Finally the test was set for the next day, and he passed it satisfactorily. He commented then to Frances, "Sometimes in life you have to take a chance."

So he wrote:

Dec. 3, 1946

Dear Mr. Haynes,

All my thanks for the things you sent. Against my tutor's advice, I took a three-hour trial exam (a N.Y. regents examination) in two hours, getting 77%. Consequently I don't think I will have any trouble with the one you sent.

The other day I listed forty-six chemical experiments that I have done in the last five or six years. I will write them up briefly in the hope that they will take care of my chemistry lab. for some time to come. Best wishes to all.

It was after this that he made the mildly ironic comment, "Well, my tutors have finally caught up with me!"

* * *

Then he worsened sharply. The bump looked like two tomatoes and he became very tired and feverish, with the fever climbing uncomfortably high. Of course he had had some fever right along. Smears showed that staphylococci were present now, and this seemed to confirm Mount's worst fear, that infection of the meninges might occur. Putnam came back from California and paid a call. He was amazed that Johnny was still alive—let alone that he was well enough to take and pass examinations on schoolwork of the year before. Literally it seemed that Putnam could not believe his eyes, and he, Traeger, and Gerson talked most of an evening, behind closed doors, while Frances and I waited nervously upstairs, and Johnny dozed.

Everybody—except Gerson—thought that we must have a prompt minor operation to avert a wide infection. That bump had become very dangerous, exuding large amounts of pus every day. Also the adjacent skin was beginning to break down, which would in time mean more ulceration. Again, though the bump was bigger, it was very soft now, and drainage might be possible. But Gerson fought like a tiger against this view. His theory was that the tumor was now *dead*, killed by the diet, and that the suppuration consisted of nothing but dead matter; the tumor was, as it were, sloughing itself slowly out of Johnny's head. Also he resolutely opposed an operation on the ground that anesthesia would be fatal. Johnny had completely won the hearts of everybody at Mrs. Seeley's. Gerson loved Johnny and wanted to save him as much as anybody, and he wanted moreover to save him in his own particular way.

So began a battle of the doctors that all but destroyed us. I have never known such strain as during that December week. The ultimate decision and responsibility rested, of course, on Frances and me. The doctors could only explain and suggest; what they did was, in the final analysis, up to us.

Finally we compromised. Traeger suggested a freezing agent instead of an injectable anesthetic; Gerson finally agreed to this, Mount agreed to operate, and everybody else agreed that even while Johnny was in Neurological he would remain on the Gerson diet. This meant weighty troubles for Frances, because the food and utensils had to be brought from the East 60's to West 168th Street every day. Johnny was, I imagine, the first patient Medical Center has ever had who was not allowed to eat from its own kitchens. What a scandalized commotion the nurses made!

Johnny's own comment after Putnam's visit was revealing. "Of course operate. The bump is poisoning my nerves." He went on: "The bump will open twice more." And as it turned out he was dead right.

So I drove him to Neurological again for what we thought would be a stay of a day or two. He stayed five weeks.

*   *   *

The operation was scheduled for the afternoon. Early that morning the bump spontaneously opened of itself, as Gerson had stubbornly predicted it would, and Mount, summoned by Johnny himself who realized exactly what was happening, did the evacuation right in his room, because there was no time to move him. Mount called me at about eleven in the morning, his voice fairly choked with joy, saying that he had successfully drained an abscess that went five centimeters into the brain beyond the table of the skull, and had got out a full cup of pus and fluid.

Now Johnny recovered with great leaps. He regained his confidence about chess, studied hard, greeted friends, loafed, teased the nurses, yawned and stretched and laughed. He passed another exam, which was a real one and which had to be done in a stipulated time, although he was interrupted over and over. His temperature was taken and the wound had to be dressed once, the telephone kept ringing and scrubwomen cleaned up his room, while he was actually at work against time.

"My goodness, Father! How can anybody be expected to pass an exam under such circumstances!" His voice was testy, but also pleased.

He was massively bandaged and every few hours, under the most elaborate procedure, he had to have a penicillin drip as the pear-shaped sac in the brain slowly closed with healthy granulation tissue, but otherwise he was more comfortable than at any other period during his illness. That horrible, ferocious bump was altogether gone. It had disappeared. Mount had sucked it out completely. Johnny's skull would be as smooth and normal as mine, except for the scars of the original incision which the hair would cover. Then next year—so we thought—when any further last remnants of the dead tumor had gone, we would put in a plate and all would be well forevermore.

Doctor after doctor came in to see Johnny, and expressed their free amazement. Miller, one of our old friends, told us that when he heard that Johnny was in hospital again, his first thought was of complete surprise that he was still living, second that he must be in coma and had been brought in for final palliative measures. And there Johnny was, sitting up in bed stoutly and arguing about the possibility of ionizing lithium hydride!

Then, after some days, the pathologist's report came in and we learned that the discharged matter showed no infection—the pus was sterile. The cultures showed no growth at all. Of course sterile abscesses are not unknown by any means; the tumor might have been cut off from its blood supply and the resulting necrotic tissue would be sterile. Even so, the sterility of the abscess seemed to be a tremendous confirmation of the Gerson theory that the tumor was, at least in part, indeed dead, and was emptying itself out as liquid. Gerson himself was dancing with delight. We kept recalling the puncture Mount had made two months before, when the bump had been immovably rigid. What a change since then! Finally came the day when Mount announced that Johnny's eyes were normal—with no papilledema at all—and that he considered the tumor to have been "arrested."

My sister was with Johnny and me when he got this news that the eyes were normal. I never knew till this moment just how anguishing was the strain that he strove so hard to conceal. He jumped bolt upright; then slowly, proudly, very slowly and proudly, he relaxed downward to the pillow, while across his face spread the most beatifically happy expression I have ever seen on a human being, and his eyes—normal eyes now—filled just to the brim with tears, but did not spill over, as he smiled with

relief, pride, and the exhaustion that comes with
release from intolerable strain.

In a second he had recovered, and was telephon-
ing his mother. He grinned. "Every telephone
really needs a bed beside it."

But there were still plenty of confusions and disappointments too. One doctor would contradict
another and then himself—because, in truth, the
circumstances were so unprecedented. They were
terrifically impressed at what had happened, but
they could not explain it or vouch for the future.
They soberly could not believe that the Gerson
regime alone had produced this effect. But when
we asked them "Would you yourself take the responsibility for taking Johnny *off* that diet, *now?*"
they all said, "No!"

We found out more and more how sickness
makes a world of its own. Johnny came to feel that
people not in hospital were representatives prac-
tically of a different race, from another planet. He
banded together with his fellow patients, fellow
prisoners, in a kind of mutual defense pact, walling
out the external world. But how he envied the
healthy outlanders! What Frances did was to give
him an efficient balance between the one world

and the other. She taught him to adjust himself to the life of the hospital while still maintaining active touch with what went on outside. It was I who was the epitome of the external world breaking in. Sometimes I was too exuberant; the contrast was too acute, and though Johnny loved my visit every evening and looked forward to it eagerly, the letdown after I left was sometimes sharp.

He still felt very strongly that he was actively participating in *Inside U.S.A.* Late in December the Book-of-the-Month Club accepted it, though I still had a good deal of writing to do. Johnny sighed. "Well, that solves the financial problem!"

A little later something gave him great happiness. He had sent a question in to "Information Please," and it was used on the program one night when I was a guest. I had asked him to listen in but he had no idea that his question was going to be used, and he went wild with excitement when he heard his name on the air. The question asked us to tell what the symbols K, K2, K., and K₂ signified. Johnny stumped us handsomely, and immediately made plans for use of the Encyclopedia Britannica he had so nicely won.

He was gay and confident. He murmured to a visitor, "The doctors are fighting among themselves now as to who cured me."

Mount let him come home for 36 hours over Christmas, penicillin drip and all, and we had a small party and a happy time. But then he had to return to the hospital because it took time for the abscess cavity to fill. This was Christmas night. I will never forget Johnny's calmness, covering over his heartbreak, as I drove him back and he limped down the long, empty corridor, and then hiked himself wearily into bed and drank some of his juices—so lonely, so alone, so unyielding, and with the hospital cold and stony and most of the nurses away for Christmas, after the warmth and lights and the presents under the tree at home. "Well, Father," he said at last, "good night."

He was not discharged till January 12. He wanted urgently to go home, but we decided to fill a small additional interval at Mrs. Seeley's; of course we were still following the diet adhesively. Frances drove him up to the hospital for a series of last dressings, and finally he was in his own comfortable small room at home again. He had been away since August, and this was February 6.

In a minute he was jumping around arranging the chemicals on the laboratory shelf. Mount came up a few days later and, venturing beyond anything he had ever said before, expressed the opinion that the tumor was "quiescent." The miracle had happened. We were wild with hope.

———

CHAPTER 4

TO THIS day, what caused Johnny's spectacular improvement during the winter is unknown. Anybody may have his guess; the plain fact is that we simply do not know. The recovery may have been due to the X-rays, the effect of which is often delayed and cumulative; to the mustard which can do unpredictable things to a body; to the fact of his youth and the growth of healthy cells despite the tumor; to mysteries in the human spirit; to the Gerson diet; or to combinations and permutations among all these. Similarly, we do not know—nobody knows—what caused the severe deterioration that came next. Something, *some*thing kicked that volcanic tumor loose again. We do not know what. All we know is that for some months Johnny was miraculously better, and then very suddenly and sorely worse again.

February, 1947, started very well. Three months before, Johnny had scarcely been able to walk. Now, though it would be an exaggeration to say that he romped all over the place, he was capable of walking half a mile or so. He veered a little to the right, his left foot was wobbly, and he needed a modicum of guidance—which of course he resolutely refused to admit—but the improvement was incontestable. Even his left hand could not have been very bad at this time, because one evening, at his insistence, I watched him give himself a hypodermic injection of liver extract on the side above the hip, an awkward place to reach. I could not possibly have done on anybody, let alone myself, what Johnny did so skillfully. He took the big syringe apart, boiled the two sections, put it together unaided, even though his fingers had little grip, inserted the needle into the ampule, drew it out, carefully tested it to eliminate the tiny bubbles, pinched up the flesh for the injection, explored to see if there were any veins nearby, and rammed the inch-and-a-half-long needle home. He sat down, grinning; and the sweat was coming out on my forehead, but not on his.

Frances had rehearsed this with him several

[127]

times, though she had a horror of doing the actual injection herself. In fact one of the reasons he did it alone was to spare her. They called it "bayonet practice."

Steadily, too, she helped him in physical activity. She taught him to be deliberate about almost every physical movement, so that he picked his stance carefully and knew exactly what he was going to do next before doing it. The stronger parts of his body must, she told him, be trained to assist those weaker; more and more his good right side must take up some of the burden from the other; she showed him how his right hand could unobtrusively assist the left. In all this two mentors were of assistance—Buddha, with his lesson BE AWARE, and, of all dissimilar personalities, Mr. La Guardia and his slogan PATIENCE AND FORTITUDE.

Johnny was vigorously interested, too, in things external. An item came in the papers to the effect that the Dutch royal family was having some sort of minor dynastic trouble. He sighed. "To think that after two thousand years of science, history still gets snarled by that sort of thing!"

Years before, at Lincoln, Johnny had met and liked a girl named Mary. Often Frances talked to him about falling in love and marrying some day; once he smiled in reply, "Very well, just throw a chemist across my path." Mary came back to his mind. Johnny was very casual about it. "Oh, by the way . . ." he began with Frances, and then recalled Mary to her attention. Frances called her mother, and then Johnny talked to Mary herself on the telephone, and they arranged to meet. Johnny felt very proud and grown up. He chuckled later. "As soon as I talked to her my temperature went up and it's taken three days to get it back to normal."

Once a strange thing happened. Frances, on whom this struggle was exacting a frightful toll, telephoned Mount to ask some question, after she had tried to get Traeger, and as the connection was made—the odds against this happening are probably a billion or more to one—she was cut in, by pure accident, on a talk Traeger and Mount themselves were having at that very instant. She never told them that she unwittingly overheard their conversation. I would like to think that this small episode shows how closely we were all held together by some invisible bond.

Everything seemed better day by day. But—by

the end of February it seemed that, after six weeks of being contained, the bump was ever so slightly beginning to bulge out again, almost imperceptibly, but definitely.

On February 18, Gerson held a demonstration. This was against our wishes. But inasmuch as we thought he had been largely instrumental in saving Johnny's life, we could not refuse permission. Gerson, absolutely sure that Johnny was saved and very proud, invited some twenty prominent New York officials and physicians to see him and six or seven other patients. He had taken color photographs of Johnny's head during the preceding autumn; we saw to it that Johnny would not have opportunity to see these, which might well have frightened him. The assembled doctors looked over his chart and all his records back to the original operation, and then examined his skull which was beautifully healed, watched him walk, tested his grip, and asked a few questions. All went smoothly and swiftly. Yet I am convinced that this event—perhaps the half-hour wait with the other Gerson patients—upset Johnny gravely. He seemed indifferent, however. Frances and I walked him home, and he stopped at my hotel for tea.

The next day, February 19, at breakfast, he had

a sharp attack of trembling, together with amnesia. He talked to Frances more or less like this:

"Is it eight o'clock in the morning or the evening?

"Where am I?

"What happened yesterday?

"I can't remember.

"Oh, yes, of course.

"It's the queerest thing—I can't remember!"

The most remarkable thing about this was that Johnny had completely forgotten the exhibition with Gerson the day before. Of course he was subconsciously blotting out from himself and obliterating what must have been a frightening experience. Absolutely all memory of the preceding day had vanished. But none of our doctors seemed much troubled by this amnesia, terrifying as it was to us. I talked to Putnam on the telephone and he said that it might well occur, as the result of temporary blood displacement, even if the tumor *were* actually dead. Mount came and made a thorough examination on February 22. The head still looked wonderfully better, though a small odd blister had developed, and, despite the amnesia, Mount was as optimistic as I ever saw him.

But the next day, though he loved Mount and

had been with him for a full hour, Johnny could not remember his visit until we jogged his memory. Again, he was trying to blot out all recollection of something that alarmed and worried him. A day or two later, feeling his head (he never looked at it), he himself could see that the bulge, though closed—we never had any leakage trouble again—was ever so slightly bigger. He turned to me with a grim voice. "Father, if it comes out again, will it be *for the last time?*"

Several times Johnny had these amnesia attacks, and Frances developed a technique for handling them. Usually the attacks came if Johnny thought that a *doctor* was worried about him. He sought to believe in each of his doctors implicitly—more than we did, perhaps. He had to, to survive. Usually his first question was, "Where am I?" and then, "What day is it?" or "What year is it?" He always—helping himself by groping toward concrete reality—strove first to *place* himself in both space and time. Then, as a rule, he would ask about Deerfield and when he could return there. Frances sought to shield him from the terror she felt herself; she would pull herself together, laugh with him, and say sharply if necessary, "Who are you?—You are John Gunther, Jr. and you know perfectly well who you are—you live *here*, and don't joke with me!" Then she would proceed and tell him what date it was, of the Christian era, and remind him how he had wanted to throw a baked potato at her the night before, and slowly progress outward in both space and time—showing him his notebook, perhaps, or something similarly concrete—until he had recovered. Usually, mercifully for her, the attacks were brief.

During the next days he talked incessantly of schoolwork and the Harvard entrance examinations. He wanted to know why, if he *had* to stay on the diet, he could not go back to Deerfield, live on the diet in the infirmary there, and resume his classes. The diet was certainly a major worry and preoccupation. Yet we could not dare to stop it. No doctor, even the most orthodox, would let us even consider taking him off the diet now. We had comforted Johnny by saying that he could quit the diet in "June," and day after day he would count the remaining weeks on his fingers. Gerson, of course, thought he might have to remain on the diet at least for a year and maybe for several years.

Medically, the situation became more puzzling

all the time. During March, Johnny was in good shape, though the bulge continued to grow very slowly and he had intermittent periods of seeming dazed and dopey and the hand and foot became somewhat worse and the lag in the mouth deepened. Also he was apt to drop things if he didn't watch his left hand. But the eyes were better. There was no recurrence of papilledema. How, we wondered, could he get better in one direction and worse in another, at one and the same time? We took him to an eye specialist who said that he did not think Johnny's eyesight would deteriorate further, and—blessed relief—that we need no longer fear blindness. But—suppose the partial numbness, the insidious creeping paralysis, should spread and so effect the right hand and foot in addition to the left? Some doctors said this was impossible. Others disagreed. If only the doctors would get together!

Straight through March and April, despite everything, Johnny worked on and on. He was utterly obsessed about getting into Harvard in the fall. But to achieve this he had to complete making up his work at Deerfield and graduate as well as pass the college entrance exams, a double task that

seemed impossible. Then on March 18 we learned
that he had caught up with his history course at
Deerfield—though he hadn't been there for eleven
months—and on April 7 he had a letter from Mr.
Boyden that gave him radiant happiness: he had
passed his English examination satisfactorily and
so was abreast of this course too. I took him to
the science room in the Public Library, where he
did some advanced work, and he proceeded to
write up no fewer than fifty-four chemistry experiments! Then Frances found out about the New
York Tutoring School, where Mr. Matthew is a
wise and considerate headmaster, and we enrolled
him there. Johnny's marks after six weeks were 90
in English, 95 in history, 95 in trig. This in a boy
with half a brain!

We made formal application for him to enter
Harvard, and in his own handwriting, now wretch-
edly uneven, he wrote the following paragraph to
answer the question, "What are your reasons for
going to college and why do you wish particularly
to come to Harvard?"

I wish to go to college primarily to complete a
sound general education and to prepare myself for
the years to come. Also, I wish to prepare myself for

research work in physical chemistry. I have chosen Harvard to atain [*sic*] these ends because I have been advised that it is the institution where I may most fruitfully attain these aims.

On April 12 he took the college board exams. It had been arranged, through the courtesy of Dean Gummere of Harvard, that he could split these into two sessions, taking the Aptitude tests at this time and the Achievement tests later. Johnny said he preferred to do both together, and who were we to say no to him? He announced, "I never felt better in my life." But it was a grueling day. We drove him down to a school near Gramercy Park early in the morning, and he had to wait almost an hour, standing mostly, in a huge milling crowd of tough, husky youngsters—many of them G.I.'s who had seen combat abroad—and then squeeze his way inch by inch into elevators packed and jammed, and run down long confusing corridors. A couple of the other boys laughed at him, and he flinched. Johnny looked so pitiably frail. His gait was lopsided, and his bandage made his pallor even more striking than it would have been otherwise. Frances kept close by. The exams lasted six and a half hours. We didn't get home till dinnertime, and

Johnny flung himself on the couch, exclaiming, "Boy, *am* I tired!" Then he eagerly snapped to the phone to compare notes with other boys who had taken the same exams.

That morning Frances had picked up a benzedrine tablet from Traeger and she asked Johnny to take it, thinking it would stimulate him during the ordeal. He had never had benzedrine and he refused, couching the refusal in a quotation from Shakespeare to the effect that he preferred the terrors of the known to those unknown.

There were other activities, too. The day he enrolled at the Tutoring School I took him to *The Best Years of Our Lives* as a kind of celebration, my sister took him to *Brigadoon,* and with Frances and me he saw matinées of *Joan of Lorraine* and *Finian's Rainbow* (wearing glasses for the first time which made him look like a juvenile professor). He kept up his acute interest in politics and he still read the news of the day with great care. I took him to shop for a lot of clothes—he had grown at least three inches during all this time—and our friends dropped in and once or twice we played chess and of course he beat me. So that I might improve my game he wrote out a series of

hints and precepts; any chess enthusiast will realize
how sound they are, and at the risk of boring the
non-chess-playing reader I will print them:

1. Always try to maintain control of as many center
   squares as possible with pawns.
2. Never bring out the queen early.
3. If winning, that is if you are ahead in material,
   exchange pieces but not pawns, so as to make
   it easier to queen a pawn.
4. If behind, exchange pawns not pieces.
5. The Kings bishop pawn is a weak pawn and
   always try to protect the KB2 square.
6. Never defend a piece by attacking another, since
   a good move by your opponent will mean that
   you will have two pieces to defend instead of
   one.
7. In a Ruy Lopez, if the opponent does not move
   the rook's pawn up, castle.
8. There's a great difference between attacking
   and just exchanging pieces.
9. Storm your opponent by attack by pawns on
   the opposite side to the one he's castled on.
10. Castling on the opposite side to your opponent
    and attacking when your position is mature will
    usually win.
11. In queening a pawn, keep the king in *front* of
    the pawn while moving.

He was very pleased and proud when, in the

middle of March, I finally finished *Inside U.S.A.*, and then he went bleak with horror when I had some mild fun writing a blurb and helping with the ads. "Oh, Father, you wouldn't!" he groaned. Later I read him a fine puff that Bill Shirer wrote. Johnny winked at me cynically. "It's swell—sure you didn't write that yourself?" He was delighted, but skeptical, about my plans for *Inside Washington*, the sequel I was outlining. I showed him the list of chapters, and half a dozen times he went over it and referred back to it, proving that I could not possibly squeeze all I wanted in my projected thirty chapters. I had insisted to him that I was never again going to produce such a monster as *Inside U.S.A.*, and that this successor would be much shorter. "Each chapter will end up two," he smiled. "I know you."

The continuing diet gave him moments of real desperation. Once he pleaded with us to invent or concoct somehow *some* sort of soft drink. We appealed to Mrs. Seeley, and she sent over a sugarless raspberry juice out of which Frances made a kind of lemonade. He sipped it greedily. One day, having finished a glass of one of the regular apple and carrot juices, he tossed the glass right out the window. It fell sixteen stories and tinkled. Johnny smiled.

The aftermath of writing a long, difficult book is almost as trying as writing the book itself, and I was very busy. I said one evening that all I really wanted in life was one full day in which there should be absolutely nothing I *had* to do. Johnny: "Try jumping off the Empire State Building."

Once Frances helped him straighten out his desk. Usually he loathed having her do this. But it was becoming increasingly difficult for him to find things, and so he was torn between resentment at being helped and the desire to get on with his work efficiently. When he did allow her to help him, he was very bossy. Once, to make things easier, she suggested that he pull out a drawer. Johnny: "Mother, I am utterly oblivious of such trivia."

* * *

Meantime our assault on what seemed to be the entire medical profession of New York City was continuing. There was no doubt—brutally, inexorably, that damned murderous bump was getting bigger, which meant that the tumor had started to grow again. To try to get a fresh over-all approach,

Frances and I went to a new neurologist, laying the
whole case before him, but aside from sympathy
he had nothing to offer, and he did not think that
the diet had anything to do with the temporary
recovery. I went to a psychiatrist to get advice as
to how matters should be broken to Johnny, in
case the paralysis should get much worse. We saw
Traeger and Gerson every week, and tried to step
up the regime with hormones. Johnny said, "Look
out or you'll make a sex maniac out of me!" It
seemed that another operation would be necessary
very soon, or the skin around the bump would
break down again, this time beyond repair. Finally,
when the bump was as big as an apple again,
Mount decided to make an exploration, and I took
Johnny up to Neurological for a two-day visit. He
was annoyed as usual at this interruption to his
studies. Mount and Miller slipped into Johnny's
room and remained there two and a quarter hours,
while other doctors and nurses constantly went in
and out. I knew the procedure could not be dangerous because I had not been asked to sign the
usual waiver of responsibility, yet I paced the corridor for those two and a quarter hours with
extreme uneasiness. Mount hoped that, as in

[141]

December, he could go in and release a lot of fluid. Anyway he was going to take a sample of what he found and the pathologist was waiting. It was done under local anesthesia and through the door I could hear Johnny grunt once or twice. The job was finished finally and then I knew it was very bad, because both Mount and Miller backed away from me in the corridor and would not talk. I went inside.

With an infinitely slow, hopeful, almost caressing gesture Johnny put his hand up to his head. "I wonder how the bump is," he murmured. A desperate charging hope shone in his eyes. Very slowly the hand went up. He felt his head tentatively, felt it again with full palm, and then, limply, helplessly, hopelessly, the hand fell down, and this was the only time I ever saw his eyes actually spill over with tears. He lay there silent for a long moment. Then he sighed, with his eyes closed. "I daresay it will take a little time to deflate."

The bump was like a stone again, and Mount had scarcely been able to evacuate a drop.

*   *   *

The next day Johnny telephoned us early. "Bring poker chips. *And* money."

I left the hospital at 4:35 that afternoon and went to work; I had been asked to review Ernie Pyle's last book for the *Herald Tribune* and was very late with it. I called Johnny at 7:30 to tell him, rather boastfully, I fear, that I had read the book and done a short review. Johnny: "Oh, Father, how could you possibly read a book in that time? What if someone reviews *Inside U.S.A.* after reading it that way?"

He asked Mount the next morning, "Did you go deep enough so that you could make me twitch?" He tried to sound cheerful, but his voice was weak and quavering. Then he asked why he had to continue on the diet if it were not doing what it was supposed to do. He added quickly, "But don't let anybody know that I'm impatient."

I took him home after a brief talk with Mount. It was all hopeless now. What Johnny had now, the pathologist said, was one of the worst of all forms of tumor, glioma multiforme; moreover the specimen Mount got showed many mitotic figures: i.e., the production of malignant cells was rapidly increasing. All neurologists are pessimists. I tried to argue with the diagnosis by pointing out how generally well Johnny still was. And indeed the doctors were astonished to the point of stupefaction that he had been able to take those college entrance exams. But Mount, so kind and decent, could offer no hope at all when we asked him what to do next. He said, "Let Johnny do exactly what he wants to do and die happy."

* * *

April 25 was a tremendous day. For some time Johnny had been working at the Tutoring School on a chemical experiment that was to end all experiments—and nearly did. Also Penfield telephoned from Montreal to say that he would be in New York and would like to see Johnny. The appointment was set for midmorning and I have never seen Johnny so choked with disappointment and indeed despair—this would mean that he would miss his precious chemistry hour and it might be weeks before he could get back to the experiment. Every moment was golden to him now. We called Montreal back and Penfield shifted his plans to fix the consultation for an earlier hour.

Then—some modicum of hope again! Penfield, with his immense authority, was greatly impressed at how comparatively well Johnny still was, and seemed to think that all was not yet lost. I showed

him what he had written on the chart the year
before; he replied, "Well, the best of us can be
mistaken." He gave the impression that Johnny
might still be saved; he urged giving up the diet,
trying the mustard again, and above all having a
second operation. He thought that the tumor *might*
conceivably have a frontier and *might* be remov-
able at this stage. Tumors in children are, he said,
unpredictable, and this one seemed to be "weep-
ing itself out" in fluid and it might not be "so
frightfully big." This was the best moment for an
operation, too, since the ulcer had healed and the
risk of infection was at a minimum. He urged
Mount to do a "whopping big job" and to risk
everything by the most drastic possible operative
procedure.

I took Johnny off to the Tutoring School and
fetched him later. He was agog with triumph and
excitement. He had done his experiment, and it
certainly had worked. I had bought him some
metallic lithium out of which he made anhydrous
lithium hydride. The stuff was, he knew, highly
inflammable. He did not know that it would burn
right through a pyrex flask, through a metal lab-

oratory desk, and through the floor. Luckily his teacher was close by.

Steadfastly, all these months, we had held to the diet, but now there seemed no point in going on. Not only, despite the diet, was the tumor growing again; it was growing faster. It would have been heartless to have continued to impose on Johnny, for no good end, a regime of such severity. He had been on the diet seven and a half months—yet the tumor was bigger now than ever before. Nevertheless we were grateful to this splendid human being, Max Gerson, for all he did for Johnny—and what he did was a lot—and we always will be. There were long consultations with Traeger and Gerson and finally we told Johnny the great news one morning and he asked suspiciously, "What strings are attached?" Only, we replied, that there ought to be a few intervening days, for weaning so to speak, when he would still avoid salt and fat. Frances and I had lunch out that day, and then we rushed to a butcher shop and bought a steak. We got home—and Johnny wasn't there! He had slipped out of the apartment unnoticed by anyone and had walked by himself to a nearby Translux to take in a newsreel. This was the first time he ever did this. Of course it was a symbol of release. That evening he ate the steak and more or less anything he liked. He kept gripping a huge piece of hard cheese in his good hand and nibbling at it hungrily.

We had a stiff fight about his going out alone. Frances usually took him to the Tutoring School, luckily not far away, and sometimes I called for him. He resented bitterly being accompanied, but we did not quite dare to let him go alone, even by taxi, because of the steep stairs at the school. It became a question of whether it was better to risk an accident—of course if he should fall and hit his skull, it would be fatal—or to damage his self-confidence by continuing to withhold permission. The matter was taken out of our hands promptly. It became clear that the operation, if it was to take place at all, must take place very soon, since the bump was now all but bursting out of the skin like a vicious cauliflower.

I took him to see *Odd Man Out* one afternoon, and his friend Mary, his cousin Judy, and Edgar Brenner came over in the evening, and I sat in his room while the children played cards in the living room, and as always it was a struggle—between letting him do just what he wanted to do, and watching carefully to see that he did not get too tired. This was a dinner party all his own, and he was immensely happy. A little later Johnny sent Mary a graduation present, and he got from her a key chain which made him inordinately proud.

We went up to Neurological—once more that drive through the crosstown traffic and then skimming northward along the river—on the evening of April 29, just a year after the first outbreak of his illness. Johnny had not, of course, been told that any major operation was impending. We left after a while and the minute we arrived back at the apartment the phone rang and it was Johnny, indignant and very frightened. A nurse had come in to give him an enema. So he knew that something more than just a test was in preparation. Then he demanded to talk to me to find out whether, if there was to be an operation, we had cleared with Gerson what kind of anesthetic would be used, since he had heard him say that anesthetics were fatal to people on his diet. (In blunt fact Johnny was under anesthesia for hour after hour, and no damage occurred at all.)

Johnny wanted to live, all right. But one of his

mild remarks the next day was, "Maybe the next world will be a pleasanter place than this."

The barber, Tony, came into Johnny's room at about 9:30 on May 1. Johnny took a look at him, remembered him from the year before, and with one swift continuous gesture reached for the phone and in an instant had Frances. He said, "The executioner is here. Oh, my! It's Tony. The guillotine. *No!* Protect me!"

Frances said, "Tell Father to throw him out of the window."

Johnny said, "Father is too polite."

He really was scared. He knew that a complete barbering job meant a *big* operation, and, more important, this must mean in turn that his condition was not good. He set up such a commotion that Tony, after plugging in his electric clippers, left the room hurriedly with a nurse. In a minute the nurse returned and said, "Dr. Garcia says that a haircut isn't necessary after all." (Of course the doctor, outside, decided that rather than frighten Johnny they would cut his hair while he was under anesthesia.)

Johnny, with a magnificently regal gesture, like the hero in a comedy of manners, swept his hand out and commanded, "A bottle of champagne for Dr. Garcia!"

He went upstairs at 1:10 P.M. and did not come down again till 7:25. We waited, knowing nothing, for almost seven hours.

As he was being wheeled out he gripped my hand very tight and said, "They're not taking me to the tenth floor, are they?" I lied. "Just for a test." I walked with the stretcher to the elevator. Johnny said, "I think I will be taking a little expedition this afternoon. So long."

The nurse was fixing up his room and moving the bed that would follow him to the operating room. "Come on, bed, get going," she said crossly.

*   *   *

He was unconscious when he came back, of course. But he did not look so bruised and swollen as after the operation the year before. The anesthetist, pacing slowly, as if measuring the steps, came along with the bed, holding aloft a pink beaker filled with fluid that dripped into Johnny's veins, holding it up at arms' length quite stiff and straight, so that he looked like an acolyte bearing aloft a torch grown pale. Mount came in after a

[150]

while, white. "I got two handfuls," was all he said.

Later he told us that the tumor was growing so fast that the blood vessel nearby was thrombosed, that the malignant mass was even invading the scalp and that despite the depth he had reached, 11 cm., he had never penetrated to healthy brain tissue at all.

* * *

I was up there early the next morning, May 2, and Johnny was not only conscious but quite talkative. The very first thing he said was, "Is all the work on the book cleared up?" Then he asked where Frances was and proceeded: "Did Clare accompany us?" This was striking, because he had not seen Mrs. Luce in several months. Then: "What have you written for Bill Shirer's book?"

He had some food and demanded, "Where is Putnam?"

I said that Putnam was out in California.

"Why doesn't he hurry the hell back? He'd let me walk!"

Then: "Where's Penfield?"

After a while he said, "Let me telephone Gerson and find out what to eat." I said that from now on he could eat absolutely anything. "Except eggs," he put in, and I still do not know the reason for this remark. At noon I said, "I'm going to run out for a minute and grab some lunch." He replied, "You never eat lunch." Frances arrived, and he exclaimed to her, "It was very unfair. They squirted my mouth full of something, and so I couldn't talk."

He felt a bandage. "That's just for decoration. Why cover it up? What in heck for?"

He said he wanted to go to the bathroom and we rang for a nurse. "You can't walk," I said. "I could crawl," he said. He added, "I have your blue pajamas on. Not the pants, though. They're prejudiced here against pajama pants."

We sat with him.

"It was very unfair," he repeated. "They sprayed me full of stuff, and I couldn't protest. I tried to protest when the barber came, but they gave me something so I couldn't talk."

Later: "What happens when I get out of this labyrinth? If I go back to Tutoring School, can I walk there by myself?"

Then: "Look at my arm. *Some* bruise!" This was where they had opened the vein for the transfusion. "They put me under general, not local," he complained again, "and so I couldn't see anything interesting!"

He dozed a while—all this within a very few hours of one of the most terrible operations a human being can undergo—and then concluded: "It was the works, all right."

Finally he said, "It's better this time than after Putnam's operation. I can see."

CHAPTER 5

SMOOTHLY, steadily, ominously, the next two weeks slipped by. The bulge disappeared entirely for an interval and was replaced by what we had prayed for for a year, a concavity. Johnny was worried, though. The bandage was too big for him to feel through. On May 4 he said, "I must ask Connie [Traeger] if Mount got out as much as Putnam did." He said to a visitor, "I wonder if the bump is still there. I'm not convinced." Also he began to inquire with great earnestness why a plate had not been inserted, which must have meant that, finally, he had given up hope that the bone would grow back of itself. He said indignantly to one friend, "If only they'd put in a plate, one of the new types of plate made of tantalum, at least I'd be able to swim and sail."

Frances, who was holding up wonderfully under a strain that had become unendurable, went off to

[154]

Florida for a brief rest, and, with my sister standing by, I got away later for a week in Virginia. There came one violent hour in the solarium before I left. Johnny was passionate and stormy. He exploded, "I'm always in a haze! I was in a haze up in school last year! The tumor must have been starting then, only nobody knew it! People kidded me about it, and it was very disagreeable. I talked it over with Steve [one of his classmates] but came to no conclusion. I'm sensitive about being kidded. I didn't like it, and I don't like it now! It wasn't my fault that I was in such a haze!"

He relaxed. "My mother and father think that anything connected with me is remarkable. These strange parents . . ."

Mount dropped in and he turned to him furiously. "Get me out of here by crack of dawn Thursday, or I'll sneak out by myself! Stop all this persecution!" He asked him to speed up his recovery by giving him electric shock therapy. Mount, with his deadpan face and the grave, warm brown eyes, was so dumfounded by Johnny's outburst that he did not know what to say. He fumbled and tried to joke. "Electricity costs a lot and we save it for our serious cases."

Then Johnny attacked my book. "You step backward to be fair too much. You ought to have more muckracking (*sic*). You beat around the bush. You should have begun the Dewey chapter with the simple statement, 'Nobody trusts Tom Dewey!' That was what you were trying to say in five thousand words. Libel lawyers? Fire the libel lawyers! You've written three books, they've all sold half a million copies, tell Cass to take it or leave it and fire the lawyers!"

But there were lighter moments. One nurse said, "I'll be just like your mother to you." He answered, "Okay, provided you don't go too far." I was being profiled by Dick Rovere of the *New Yorker*, and I brought Rovere up there one day and Johnny looked him over and said, "So you're the hatchet man." Later he told me, "Well, I hope he digs up a lot of dirt about you." I said Rovere was a fine fellow. Johnny: "Wait till you see what kind of piece he writes."

"Did you sleep well?" the nurse asked one morning.

"Like an octopus."

This is, I believe, the last letter he ever wrote. He had pleaded with his mother to get away for her brief holiday, and while she was in Florida he telephoned her a couple of times and then wrote this:

DEAR MOTHER,

Today is the last day at this——hospital! Thanks for your letters, and be sure to remember the words of wisdom which I tried to impart to you a few minutes before your departure. I feel fine but seem to be struck with a most monstrous attack of lazyness. What a job it is for my poor nurse to get me up in the morning! I've gotten into an awful habit of drinking coffee in the morning, and find it necessary to keep me awake—at least enough so I don't fall asleep and drown in the bathtub!

It made me happy to hear that you will be returning soon. In a week or so I will go back to Deerfield to take the exams, and to say hello and goodbye!

O! How wonderful food is again! Bacon and eggs! salt! steaks! How I eat! mushrooms! last night I played poker with some fellow patients!—great fun! I've almost finished the English anthology which we were reading

                    lots of love and kisses

                                        JOHNNY

He was cleared to go home on May 15, only two weeks after the operation. So for the last time Johnny checked out of Neurological. He ended the experience with a wry wisecrack. We marched out and I said the hospital knew us so well by this time that they sent the bill by mail. Johnny jibed: "You mean by parcel post."

* * *

What a lot goes into a life, into a brain—all that the fragile shell of cranium holds! Usually the size of a skull, and the brain concealed within, is an index of mental capacity. Johnny's brain, we learned later, weighed two thousand grams. The average for a normal, fully-grown male is fifteen hundred. The largest male brain ever known weighed 2,222.

* * *

He remained pretty well, but now it became increasingly difficult for him to fix his belt or shoe-laces. He was too proud to admit this, and Marie, our admirable housekeeper, helped him to put on his shoes one morning. Johnny said, "I'm only giving way to your maternal instinct." Carl, our old elevator man, wept once when he saw how warped his face was and how difficult it was for him to walk. Johnny said to him coolly, "I haven't had any chance to exercise, and so my foot is tired."

Marie told me of another colloquy. A school-mate whom he had not seen for years called up.

JOHNNY: "I should warn you that my head is bandaged because I have a brain tumor."

BOY: "I've never known anybody with a brain tumor."

JOHNNY: "You know me."

BOY: "What's it like?"

JOHNNY: "I've been lucky. I have no pain, and there has been no impairment of my faculties."

The boy came over that afternoon, and Johnny cleaned him up in a game of chess.

The effort to pretend that the tumor was nothing cost him dearly; the price of his invincible fight was great fatigue. It took a miserable lot out of him to pretend to ignore what he must have now known to be the truth, that he wasn't getting any better. The faraway look was in his eyes more often now. But it was impossible for us not to support his optimism, because any discouragement would have been a crushing blow. All he had now was his will to live. We had to keep that up at any cost. The cord of life was wearing very thin, and if we took away hope, it would be bound to snap.

After a struggle one morning he gave up trying to tie his tie, and things would drop out of his left hand more frequently. "My left hand is a mess." The hand cupped sharply and he looked frightened. "The nerves are crazy in this left hand. I can't get it open."

He always loved to joke with me about my size. I said one day that I was tired enough to stay asleep until I starved to death. Reply: "That would take quite some time, Father." I had a massage and reported that I had lost some weight. Comment: "How much did the masseur lose?"

He read the papers carefully and with Frances listened to every important broadcast. He said, "The reason why the Republicans don't offend and oppose those Southern Democrats is because they may need their help some day." Some friends talked once about the great vitality of the United States. He asked, "But may not vitality end in smugness? Isn't it possible, too, that vitality could express itself in reaction, in the wrong direction?" He turned to me. "In Volume Two, hit them hard, Father!" I can tell you all right whom he meant by "them"—anybody cheap, anybody shoddy or vulgar, anybody selfish and corrupt, anybody on-the-make or feathering his nest in the name of false principle.

He dropped a pill.

"Is it still all right?" I asked stupidly enough, reaching for it.

"It will be if you pick it up off the floor."

On Sunday mornings Frances read to him from the Hebrew Bible, the Christian Gospels, the Hindu scriptures, Confucius, and other eastern sages. One of the last things he read was the Psalms. I read to him, too, though not so much. One of the books he was going through for English was a poetry anthology; he would look bored or turn away whenever we chanced on a poem about Death.

One day came an unbearably moving moment when he announced, as if casually, that perhaps he was having the bump for *us*!

The phone rang on May 25 and Mr. Boyden's cheerful, assured voice came through. "I've gone through Johnny's papers and examinations," he said. "You know he did extra work in his freshman year and has some surplus credits. He has caught up to his class in everything except one examination, and we are going to give him a diploma. This isn't a favor. It is Johnny's right. Come up next week, and he will graduate with his class."

Johnny yawned and tried to look casual, and we all burst into tears.

* * *

We drove to Deerfield on May 27, and Johnny graduated on June 4, though he had not been to school for fourteen months. The days passed in a proud procession, and I think probably it was the happiest week of his life.

It seemed chilly when we started, and Johnny, as always extracting compensation out of any ill fortune, said, "Well, at least we don't have a heat wave." We passed through Hartford and he asked, "Were you here when you did your research?— I wouldn't dream of asking how long you stayed, probably half an hour." I was full of nerves as we got near Deerfield with its stiff old houses and great fanlike elms, and impatiently I asked him if I had overshot the side road and did he recognize any landmarks. He replied gently, "You know I don't see well out of my left eye."

Then without the slightest self-consciousness he took his place in his class. He sat between old

[162]

friends in the dining hall (the instructors had
warned them) and Frances whispered that they
should inconspicuously cut his meat if necessary.
The boys stared at him for a second as if he were
a ghost—of course his hair had not grown back
fully after the last operation and he wore a white
turban—and then accepted his appearance without
question.

Every evening after dinner an informal ceremony takes place at Deerfield which is one of the
distinguishing marks of this magnificent school;
each boy from Freshman to Senior meets with Mr.
Boyden, and the roll of the entire school is called.
The boys are heaped together on the floor. Usually
there is a casualty or two—some youngsters hurt
in a football game—for whom there are big leather
chairs. Johnny eased himself into one of these, and
his name was called in the roll exactly as if he had
never been absent for a moment. Then he limped
slowly and proudly to the Senior Dorm where he
would have been living this past year, and looked
at what should have been his room with a piercing
yearning. Boys were moving back and forth in the
orderly bustle that precedes commencement.
Johnny had the attitude of one who is both a par-

ticipant in and a spectator of a great event. Mr. Boyden crept up to us and asked if we were sure he would not get too tired. Then he joined calmly in a bull session.

It was decided that he should sleep in the infirmary—a building he knew only too exasperatingly well. The next morning we came to pick him up at what we thought was a reasonable hour. But he had left the building before eight, alone, and was at that moment taking the final exam in chemistry! He passed it B Minus—though he had never taken a regular chemistry course in his life.

Later that day I bumped into him accidentally on the bright sunlit grass as he dragged himself from behind a hedge in shadow. His left shoulder sagged; his arm hung almost useless; his mouth was twisted with effort; the left side of his lip sank down; his eyes were filmy; he was happy. "Oh, pardon me, sir," Johnny said. He had not recognized me, and thought that I was some master he did not know.

Everybody tried hard to keep him from being too active. But he said, "Walking around this way helps the wound heal." Frances told him to sit around in the sun—how they both loved the sun!

—and get brown and he answered, "All you are interested in, Mother, is my color!" When he had trouble with knife and fork one evening, he told her in exquisite parody of what she often said, "Be patient. Believe in calmness and Nirvana." It was a lovely day the next day and Johnny spent an hour learning some calculus from a fellow student. He worked out the equations on the bottom of a paper plate during a picnic lunch in the soft grass. Frances remonstrated that he might be getting tired. He replied briefly, "There's no future to just sitting."

The day before graduation was strenuous, with a lunch for the parents at noon and then a baseball game which Johnny watched with serious interest for about four innings. The dress-up banquet that night, to celebrate among other things Mr. Boyden's forty-fifth year as headmaster, lasted three hours; Johnny did not miss a minute of it. He tramped across the lawn afterward, with his classmate Henry Eisner holding his hand, for the off-the-record talk Mr. Boyden gives each graduating class. Then the class, standing under the trees in a night grown chilly, serenaded the Boydens on the front porch. Johnny, on the outskirts of the massed

pack of boys, looked suddenly exhausted, and I
slipped away from the adults to join him inconspicuously, standing just behind him. He did not
mind, though as a rule he loathed having us anywhere near him at school. I was afraid he might
fall. Then I heard his light, silvery tenor chime in
with the other voices. The song floated across the
lawn and echoed back. We hiked to the infirmary
and Johnny ran into a classmate who had won an
award. "Congratulations!" he snapped briskly.

The next morning the boys assembled early for
the quarter-mile walk to the white-frame Deerfield
church, arranging themselves four abreast in order
of their height. I did not think Johnny could manage such a march. He shook us off and disappeared.
The procedure is that the boys, reaching the
church, line up behind the pews, and then walk
one by one down the center aisle, as each name is
called. Mr. Flynt, the president of the board of
trustees, then shakes hands with each boy, giving
him his diploma in the left hand. We explained
that Johnny might not be able to grasp the smooth
roll of diploma with his left fingers, and asked Mr.
Flynt to try to slip it into the right hand instead.
The boys began to march in slowly, and though

Johnny should have been conspicuous with his white bandage, we did not see him and I was in an agony fearing that he had fallen out. Mr. Boyden, sweeping the assembly with his all-embracing sharp affectionate glance, caught Frances's eye and nodded to her reassuringly. One by one the names were called out, and each boy disassociated himself from the solid group and marched forward alone. The call was alphabetical, and by the time the G's were reached we were limp with suspense, since we did not know for sure that Johnny had even got into the church. As each boy passed down the aisle, there was applause, perfunctory for some, pronounced for others. Gaines, Gillespie, Goodwin, Griffin, Gunther. Slowly, very slowly, Johnny stepped out of the mass of his fellows and trod by us, carefully keeping in the exact center of the long aisle, looking neither to the left nor the right, but straight ahead, fixedly, with the white bandage flashing in the light through the high windows, his chin up, carefully, not faltering, steady, but slowly, so very slowly. The applause began and then rose and the applause became a storm, as every single person in that old church became whipped up, tight and tense, to see if he would

[167]

make it. The applause became a thunder, it rose and soared and banged, when Johnny finally reached the pulpit. Mr. Flynt carefully tried to put the diploma in his right hand, as planned. Firmly Johnny took it from right hand to left, as was proper, and while the whole audience rocked now with release from tension, and was still wildly, thunderously applauding, he passed around to the side and, not seeing us, reached his place among his friends.

That evening we talked of Harvard. Some of the boys were getting their admission notices, and Johnny, now that he had actually been graduated, wondered when his would come. He was impatient. He had a great sense of the passage of time.

Everything that Johnny suffered was in a sense repaid by the few heroic moments of that walk down the center aisle of that church. This was his triumph and indomitable summation. Nobody who saw it will ever forget it, or be able to forget the sublime strength of will and character it took.

* * *

Back in New York we pressed on ceaselessly with medical affairs. It was simply impossible to

let this child die. The bump, that criminal ma-
rauder, was growing out again, not with so much
pace as before, but it was harder and more tense.
We had long ago ceased to think of expense; I
was very heavily in debt, but my book would make
it up, I hoped. We gathered ourselves for a final
assault. Mrs. Albert D. Lasker gave me some special leads, particularly in regard to the use of hormones. Traeger went to a convention at Atlantic
City, and there talked with every brain man, he
could find. A chase began like that of the year
before, with calls to specialists in little-known
procedures in Los Angeles, Chicago, San Francisco,
and the University of Michigan. We heard of an
extremely orthodox man on ordinary cancers who,
after he had given up hopeless cases, found the
patients responding to a therapy based on estrogen.
Frances and I talked to Craver at Memorial, and
after a long conversation in which nothing much
developed he popped out at the last moment, as
doctors will, with news of a folic acid derivative
that just conceivably might be useful. As always,
the great obstacle was that whatever the effect of
new medicaments on other types of tumor, they
had never been tried on the brain, and experimen-

tation in this area was, to say the least, risky. Still, even more so than before, there was nothing to lose. One morning Johnny himself tore a clipping from the Sunday *Times* and—mutely—showed it to Frances. The appeal in his eyes was terrible.

It described an experimental treatment for cancer and leukemia announced in Chicago; I talked to the young physicians handling it, but there seemed little possibility of applying their technique to a brain condition. Then I went out to Montefiore for a long conversation with Davidoff, a celebrated neurosurgeon to whom I tried to outline the entire case. I asked him flatly if he had *ever* known a glioma multiforme to be cured. He hesitantly adduced recessions, but not cures. How long, I wanted to know, had the longest case in his experience *lasted*. Four years, he replied. Then he talked wisely about the mysteries of the body, "miracles" such as those at Lourdes, and the inscrutable quality of some diseases, and suggested that, since diagnosis was so important, we should send Johnny's slides to two supreme pathologists, one in Illinois, one in Connecticut, for a final verdict. I asked him to what he attributed the improvement Johnny had indubitably shown during the spring. He was convinced that the diet could have had nothing whatever to do with it, and that almost certainly—this was a surprise—the amelioration had been the consequence of last year's X-rays. He told me some frightening and marvelous things about X-rays. Yet I had conferred with the radiologist at Neurological when Johnny was better and the radiologist had disclaimed any credit for the improvement. Moreover all our doctors had forbidden more X-ray. But, Davidoff said, there were million-volt machines that, so to speak, skipped the scalp, and might conceivably be useful. He didn't give us any hope, but he suggested that we have a try.

People may ask if it would not have been better if we had had fewer doctors and less treatment. Perhaps we tried to do too much. But Johnny loved life desperately and we loved him desperately and it was our duty to try absolutely everything and keep him alive as long as possible. Might he have had a better chance to live with less attention? No. That tumor, untended and unchecked, would simply have eaten away all his brain. Always we thought that, if only we could maintain life somehow, some extraordinary *new* cure might be dis-

covered. We thought of boys who died of strep-
tococcus infections just before sulfa came into use.
Our decisions were almost always dictated by
successive emergencies, with one delicate consideration poised against another, and they were not
taken lightly, I can assure you.

Now Johnny was steadily becoming more aware
of how sick he was. Did he know he was going to
die? I leave it to the evidence of his own conversation. He was lethargic and sometimes testy during
these last declining weeks. He demanded several
times to know the result of Mount's last laboratory
report; we had to equivocate, of course. He demanded to know then, "If Mount's report is satisfactory, why is further treatment still necessary?"
He demanded to know, was the bump bigger, or
smaller, or what, and was it fluctuant or not. I
said, "Leave it to the doctors' judgment."

Johnny: "Their what?"

One morning he said to the technician who took
his blood count, "What's the point of going on
with this? What does it all lead to?"

He had to take a lot of yeast and, finding the
taste offensive, he took to putting it in capsules.
I protested at how laborious this was. He answered
shortly, "Occupational therapy."

We decided finally to repeat each factor that might have led to the earlier improvement—X-ray, mustard, diet, and, conceivably, another operation if he could stand it. We chose to begin with the mustard, largely on Penfield's urging and also because we remembered how strikingly he had picked up after the first mustard the year before. And certainly HN_2 has remarkable effects in shrinking some types of tumor. After the mustard we would try X-ray and finally, late in summer, go back to Gerson. This order was prompted in part by consideration of the strain and discomfort to Johnny; X-ray and diet were the hardest things for him to bear, so we would do these last.

Burchenal wanted to hospitalize Johnny this time, and so on June 12 he went to Memorial. He swam through this experience, which was short, quite easily. There were no ill effects. But so far as we know, no good effects either. Johnny was alarmed, though. A merciful doctor who found a way to inject the mustard in tiny veins on the back of his hand, so that there was none of the usual difficulty getting into his arms, told him, "You're doing fine."

Johnny: "So far."

This doctor, Ursalof, was so impressed by his

[173]

agreeableness that he called him, with affectionate mild irony, "Mr. Aggressive." Johnny joked with his mother about why he had to be in the hospital at all and said, "Oh, you've just succumbed to Burchenal's charm, as usual." He told one of the nurses, "I'm here only to get the last little lump of that tumor out, the very last small *little* bit." But when he vomited the first evening he grunted, "Puking? That's the least of my worries!" One afternoon he asked briefly, "Is there a crematorium handy here?"

We came home, and again Johnny was distinctly better; he had two or three splendid hearty days. I took him to see *Great Expectations* and he loved it. Walter Duranty, bless him, dropped in for two long afternoons and enchanted Johnny with his conversation, making him laugh almost till he cried with anecdotes of his own schooldays at Harrow and how he had played hooky to see the Grand Prix in Paris, and with questions like, "Well, my boy, and what do you think of women?"

Later there was some dialogue about education. Johnny said, "Schools in this country make you callous and cynical."

The doctors thought it would be safe in a few

days for him to go to the country, and Frances went up to Madison to get the house ready; I spent most of the last ten days with him alone. We knew he would not be able to manage any stairs, and so Frances converted a downstairs room to his use, painting it a bright robin's egg blue, hanging cheerful curtains, and putting in order his books and chemicals. I tried, at her suggestion, to work out some plausible arrangement with Dean Gummere whereby we could tell Johnny he had been admitted to Harvard even if he could not go. This good and decent man co-operated beautifully but, as it worked out, too late.

Johnny was busy meantime in several fields. Mrs. Luce sent him du Noüy's book, *Human Destiny*, and he read it with close interest; his final comment was, "Well, he presents a case." Then she gave him another book, the name of which I have forgotten, an outright tract. Johnny said, "It says just what du Noüy says, but not so well. You distrust it because it's propagandistic." Then after a pause, "I hate unscientific books!"

Putnam returned to town, and he came to see Johnny, with Traeger in consultation, on Monday, June 23. Now we had another surprise. Putnam—

I hope I am not foreshortening too bluntly the impression his subtle, sensitive mind gave out—was a slight shade optimistic. Traeger alluded to the bump, saying, "Nobody knows what's inside that thing!" and Putnam's reply was a shrug. "We must always keep in mind that this is a very peculiar tumor indeed." Johnny was marvelously amusing with both doctors, teasing them, anticipating their tests and questions, and trying to talk them into things. Putnam told him where he could get something he wanted badly and that was hard to find, test tubes of transparent fused quartz. When Johnny had skidded from the room I said, "The tumor did incontestably disappear for a while. *Where was it?*" We talked all over and around this again and again. Putnam said that he had, out of several thousand cases, known two, just two, glioma multiformes that *had* recovered. But I did not pursue the subject further when it became apparent that these two persons had not survived with much to live with. Finally, like Davidoff, Putnam strongly urged more X-ray treatments as soon as Johnny was well enough to take them. "But why," I protested, "didn't you insist on X-rays before?" Because, he explained, the broken-down

skin of the scalp, which could never have withstood
X-rays, had been removed by Mount's operation;
now, because Mount had pulled the flap together,
there was tougher, healthier skin to deal with.
Traeger listened closely and said little. A fleeting
idea crossed my mind: that the sturdiness of this
new scalp might be forcing the tumor inward,
which would account for the recent accentuation
of the paralysis. Still, despite everything, Johnny's
condition was pretty good. Putnam's last word was,
"Let's keep on struggling."

Johnny came out of his room and walked the
two doctors to the door. His goodbye to Putnam
was, "So long. You didn't bang me up as much
as Mount did. Thanks for finding out where I
can get that quartz."

Johnny caught the reflection of Putnam's good
spirits, and we had a happy dinner. "Father,"
Johnny said, "this is better than the Colony." He
called up his mother in Madison with exuberant
glee; she asked, "How do you feel?" and he
answered, "Great!" Then he telephoned Mary but
her line was busy for a while. He shut the door
so as to talk to her privately from my room. When
he couldn't get the connection the second or

third time he came out and murmured with great zest, "God damn it!"

* * *

Came an awful morning, on June 27, when Johnny turned to me across the breakfast table and spoke as if very casually:

"Where's Mother keeping herself these days?"

Then he felt the bump. He wasn't bandaged now. His hand played on it, shocked. "What on earth is that?" I stared at him. Then:

"How long has this been going on?

"What year was I at Deerfield?

"What day is this?

"What are these pills for?

"Where was I last week?"

Then he had a short, sharp attack of shivers.

* * *

He hated to be helped. That evening at dinner, when his memory was quite right again, he tried out a system whereby he would stick his fork up the sleeve of his jacket, pinning it that way with the elbow bent, so that with his whole arm he could hold the tines straight. "I *will* learn to manage this!" he ground out between his teeth.

[178]

We were very relaxed later. But he was chilly. For some time now it had been obvious that the temperature-regulating mechanism in his brain was deranged slightly. He talked vividly about Harvard, and then about getting a plate for his skull. "By that time my hair will have grown back, too."

He was very drowsy the next day. His grip on life was lessening, though I did not yet think in terms of emergency. On the twenty-eighth he slept most of the afternoon, but, after all, he had been taking naps during most of the course of his illness, and I was not alarmed. He said that evening, with his usual careful choice of language, "Father, I hope you don't mind my being somewhat glum."

I called Frances in Madison and told her I thought she ought to come in, though there seemed no real reason for her to make a troublesome trip inasmuch as I would be driving him out in a day or two. This was a Saturday night and we planned to go to Madison on Tuesday. Frances said she would wait to hear from me Sunday at noon. I felt a persistent apprehension now, very deep and pervasive, though there seemed to be no new specific reason for it. Johnny was much better on

Sunday, but even so I called Frances again and insisted that she return, and she arrived that evening.

In actual fact this Sunday was one of the best days he had had since he first became ill. He was exhilarated and very close to me; he kept following me around and once or twice grabbed my hand. I almost picked up the phone to catch Frances and tell her that it would not be necessary to come in after all. Johnny talked about many things. We discussed Sinclair Lewis and I told him about the ups and downs in the life of an artist, of the deep, perplexing downdrafts a writer may have. He read in *A Subtreasury of American Humor*, chuckling, and then, something he didn't often do, he read aloud to me, picking up the small-type anecdotes from a recent *Reader's Digest*. One, it happened, was about a hanging. Then he said, "Inasmuch as we'll go to the country Tuesday, I think I'll pack."

But steadily he kept returning to me, as if he wanted to be particularly intimate and affectionate. All his sweetness, his remarkable goodness and pure, ineffable niceness, seemed to be bursting out of him all this day. By mid-evening, not at all tired, he had finished a neat job of packing his books and things for the summer. Here is what he chose:

> *Atomic Physics*, by the physics staff of the University of Pittsburgh
> *Einstein's Theory of Relativity*, by the Liebers, and their book on the theory of groups
> "A Poet in the Making," by Laurence Housman, clipped by Frances out of the *Atlantic Monthly*
> Some lab manuals, folders, a dictionary, the Deerfield Roster
> *Outline of First Year College Chemistry* and another chemistry text
> Beginning German
> Sprague's *Plane Trigonometry*
> *What Is Life?* by Schroedinger

Frances arrived; they had not seen each other for ten days. I had already put him to bed but he was not asleep. "Hello, Mother, I'm so glad you've come back!" he all but shouted. They talked and laughed about everything. He woke once during the night and again they talked and he whistled one lullaby.

* * *

The next morning, Monday, June 30, she gave him a bath and washed his hair. He seemed too

tired to get out of the tub. I called for him early
and took him to Memorial for a blood test and a
last clearance before his trip to the country. Nobody at the hospital saw anything amiss, though
over the phone during the weekend I had talked to
Burchenal, and he was worried that Johnny's fatigue might be caused by mounting intercranial
pressure. Most of the Memorial doctors are scientists
rather than physicians as such. Nobody at all saw
any marked change or reason for alarm; there
was certainly no intimation of impending disaster.
But Johnny could hardly drag his left foot out of
the car. Normally he would have scorned a wheel-
chair but today he welcomed it when I had the
sudden idea to get one for the long trip down the
corridor, and after a while, sitting there, he spoke
suddenly. "Say, have they got a thing around
here I could lie down on?" We had a lengthy
wait. The heat and the smell of the laboratory
animals disturbed him. There was another tedious
wait while Burchenal filled out a prescription for
some medicine we might need in the country. In
the car, returning home, Johnny seemed hazy, and
he asked whether or not Mary had come to lunch,
and how his pet turtles were, one of which had

[182]

died recently. But back home he was all right again, though tired.

Some things I do not understand. Traeger, it happened, was spending a brief holiday with Sinclair Lewis at Williamstown. He got on a plane suddenly and returned to New York, explaining to Lewis, as I heard it from Lewis himself later, "I have a feeling that Johnny Gunther will die this weekend."

I had a luncheon engagement at the Alrae with Walter Duranty and I walked over there. We were ordering coffee when I was called to the phone.

Frances said, "Johnny has a headache. Here, talk to him. He wants to talk to you."

I asked, "Have you much of a headache, Johnny?"

"And how! I've just called Connie to send over some morphine."

Frances was very much alarmed but she did not want to alarm me; she said there was no need to hurry over, but that I might call Traeger to see if the morphine was on the way. I had no premonition at this point, so far as I know, but I told Walter that if he would excuse me I would

run along without coffee, since Johnny wasn't
so well. This was the first time since the period
before the first operation, fifteen months before,
that he had had severe pain. I called Traeger, and
a drug for arresting pain, not morphine but a
caffeine derivative, was indeed en route. Then I
hiked back to our apartment. Johnny had taken a
caffeine pill and then vomited propulsively, something we had always been told to watch for and
which had never occurred before. I gave him another pill and this one stayed down. Frances told
me what had happened. To her, he had said little.
But she overheard him on the phone to Traeger
and was drastically shocked by the sharp command
in his voice, "Send some morphine, *quick!*" It
was now about 2:35 P.M. Frances wanted me to
move him from my room, where he had telephoned, back to his own room. But he seemed
too exhausted to walk and he was limp and very
heavy. I had a quick thought and lied. "Oh, by
the way, Johnny, your papers came through from
Harvard today!" He replied, "Am I admitted?"
and then yawned in a relaxed, superior, lazy sort
of way. Frances had been reading him *Arrow-*

smith. The book lay there beside the bed, open like a broom.

Still I was not really worried. We had gone through so many crises seemingly much worse. Marie made coffee and Frances and I sipped it in the living room. I had an appointment with my dentist at four, and had no thought of calling it off; I told Frances that I'd be back at five or so. Then I went in to say "So long" to Johnny and I did not like the way he looked. He was very pale indeed, and the skin was cold and moist. I called Traeger at once but even then it did not occur to me that the end had come. In fact, Traeger had stepped out and when his nurse asked me if he should be located and should come over right away, I said that I didn't think this was necessary but that I hoped he would drop in late in the afternoon, perhaps at six when he finished work. Then I went back and looked at Johnny again. He said to Frances something she could not quite make out and then something about Mr. Haynes at Deerfield, and Marie told me later that when she peeked into the room while we were having coffee, he murmured loudly "Mother" and then "Father." I walked to the

phone and called off my appointment with my
dentist and the next second Traeger called back
and I said, "Come right over—hurry!"

He stayed with Johnny a brief moment and
took me aside, pale and with a stern expression.
"He's dying. Shall we do anything or not?"

Apparently Johnny had had a cerebral hemorrhage. That tumor had eroded a blood vessel. Of
all the doctors who sketched to us so many times
how the end might come, none had ever suggested that it might be this.

The rest of the afternoon is full of harsh con-
flicting lines and shadows. Traeger called Mount
and another doctor and Mount called back and
promised to rush down from Medical Center; he
had just been stepping out of his office for his
summer holiday. He arrived—I do not know how
he could possibly have done it that fast, all the
way from the George Washington Bridge and
across town—before five. Johnny recognized him,
which was remarkable. But once more, and for
the last time, I knew everything from a doctor's
face. Mount went absolutely black and white, in
blotches like a woodcut on soft paper. He took
Frances across the room and whispered, "I'm afraid

Johnny's condition has become very much worse since Dr. Traeger telephoned." To me he said, "In my whole experience I have never known a tumor of such fierceness and rigidity."

The ambulance men came and we moved Johnny to a nearby hospital rather than Neurological, since Mount did not think that he could survive more than a very brief trip, and the hospital was just around the corner. Everything went wrong. First there were laborious and cruel negotiations on the phone. It was as if the whole fabric of our surroundings and even the most commonplace things had broken at last under this unendurably brutal strain, as if nothing at all would work, as if everything had been torn apart. It was a kind of revolt both of nature and the animate. The emergency door was locked at the hospital; its phone switchboard went to pieces crazily; a helpless nurse did not know what to do about anything; one of the attendants downstairs was hysterical; at the end, the taxi driver who took us back reeled and drove like someone very drunk, which indeed he was.

Johnny went under oxygen, of course; he was given every known medicament that could possibly help, and a youthful doctor explored, as always with difficulty, the veins in his leg for the glucose infusion and transfusion. We got to the hospital at a little after six. Frances and I sat with Johnny or paced the hall or talked on an open terrace at the end of the corridor for a series of long, vacant hours. It was a very hot, clear, dark night. Johnny slept on his side, restfully. He never regained consciousness. He died absolutely without fear, and without pain, and without knowing that he was going to die.

At a few minutes to eleven we thought we ought to go into his room; we had stepped out on the balcony for a brief second, and presently, with infinite depth, very slowly and at spaced intervals, three great quivering gasps came out of him. He had regained color just before; he had some final essential spark of animation; he was still fighting. But now these shatteringly deep breaths, arising from something so deep down that his whole body shook and trembled, told us their irrevocable message. Someone started ringing an emergency bell. After all those months of doctors and doctors and doctors, it happened that no doctor was there at that precise moment.

Not that they could have done anything. Traeger had just gone home, and he came back of course. Another doctor was in the internes' room, and he slipped up briskly. All the doctors!—helpless flies now, climbing across the granite face of death.

Johnny died at 11:02 P.M. Frances reached for him through the ugly, transparent, raincoat-like curtain of the oxygen machine. I felt his arms, cupping my hands around them, and the warmth gradually left them, receding very slowly upward from his hands. For a long time some warmth remained. Then little by little the life-color left his face, his lips became blue, and his hands were cold. What is life? It departs covertly. Like a thief Death took him.

Aftermath

ALL that is left of a life! There Johnny was, so pale, so slim and handsome, in the tweed suit with a spot on the lapel, he always had a spot on his lapel, and a bright striped necktie—with what valor he struggled to tie that necktie in the last hopeless weeks—here he lay placidly in the small chapel full of flowers, with his face sweet and composed and without a trace, not an iota, of struggle or pain, and we said goodbye to him, Frances and I and a clustering group of friends. We said goodbye. But to anybody who ever knew him, he is still alive. I do not mean merely that he lives in both of us or in the trees at Deerfield or in anything he touched truly, but that the influence, the impact, of a heroic personality continues to exert itself long after mortal bonds are snapped. Johnny transmits permanently something of what he was, since the fabric of the universe is continuous and eternal.

People tell us that that brief noontime service was something they will never forget. Frances is Jewish and at her suggestion we had a double ceremony; her friend Rabbi Newman and Mr. Neale of the Unitarian church joined to conduct the service, and in his magnificent voice Rabbi Newman read Johnny's own short "Unbeliever's Prayer." Most of our close friends had scattered for the summer; their telegrams and letters poured in, but they could not get back in time. Mr. Boyden and Mr. Nichols came down from Deerfield; Mr. Matthew came from the Tutoring School and Mr. Weaver from Connecticut and Dr. Miller from Neurological and Dr. Gerson; Mary was there, and several schoolmates, and a throng of adult friends; even his dentist came. And the flowers— I have never seen such flowers! They made a foaming loveliness, many of them scarlet and white, crimson and white: I sent bright scarlet carnations because these were Frances's favorite flower; then we saw the masses of pink and blue and tawny flowers, and a spray of golden flowers tall and spreading like a glowing bush.

When Johnny died, nature took note. There were violent squalls of hot wind, the apartment building rattled that night, and the windows shook in their steel casements. But the day of the funeral was a wonderful peaceful day, warm and with the sky very blue, clear and high and without a cloud. Frances and I drove up alone to Ferncliff and on the way back the wind came up sharply, keen and cutting but cool in the brilliant sunshine; we drove along the Hudson where we had driven with Johnny so many times, and the snapping wind under a calm sun whipped it into fresh ridges—they looked like sharp white icecaps dancing across the majestic avenue of the river.

The whys of this story, why Johnny should have been struck just in that part of him that would have been most fruitful, why his clock should have broken just at this particular time in his life, the why above all whys which is why any child should die, the whys and wherefores of the celestial bookkeeping involved, if any, I will not go into here. There are other criteria for measuring a life as well as its duration—quality, intensity. But for us there is no compensation, except that we can go to him though he cannot come to us. For

others, I would say that it was his spirit, and
only his spirit, that kept him invincibly alive
against such dreadful obstacles for so long—this
is the central pith and substance of what I am trying to write, as a mournful tribute not only to
Johnny but to the power, the wealth, the uncon-
querable beauty of the human spirit, will, and
soul.

* * *

Letters came in like an avalanche, until we
counted them by the hundred. He was just a boy
of seventeen; yet what an imprint he made on
anybody who ever met him! There were condo-
lences from camp counselors who had not seen him
for years and from a barber in a downtown hotel;
from the Negro elevator boy in my office building
and the proprietors of a friendly restaurant on
Madison Avenue; from playwrights, judges, politi-
cians, old Chicago friends whom we had not seen
in years, from teachers and doctors and newspaper
folk, old schoolmates, several of those who had
seen him graduate at Deerfield, movie people,
poets, acquaintances from faroff days in Vienna,
physicists, his godfather in Washington, the door-

men at our apartment building, refugees from Europe, his devoted governess Milla, scores of writers, and above all people who had never met him or us—parents whose sons had also died.

Of these hundreds of communications I will give only three, all from doctors.

DEAR JOHN,

Word has just reached me of poor Johnny's death— "He hath outsoared the shadow of our night"—what a gallant soul, and what an unfulfilled promise! The fact that this was to be expected makes it no easier to bear, and I hope that you and Mrs. Gunther know that you have all my deepest sympathy. Now I suppose we shall never know whether lithium anhydride will ionize in liquid ammonia—nor what ecstasies and sorrows might have befallen Johnny had he lived.

I wish I might have been of more help.

Sincerely and cordially,
TRACY PUTNAM

DEAR MR. AND MRS. GUNTHER:

What a heroic battle Johnny fought! A gallant spirit like his cannot be destroyed by a mechanical defect in the body which was given him.

Knowing him and thinking of his stubborn refusal to accept defeat makes me believe that that spirit will live on. For such there must be an immortality which we who tinker at the body may guess at but not understand.

[194]

You two, by your restless effort, kept him alive a year longer than should be expected. You could have done no more. It was worth while.

Sincerely,
WILDER PENFIELD

LIKE EVERYONE WHO KNEW JOHNNY I HAD A SPECIAL PLACE IN MY HEART FOR HIM HE WAS THE MOST GALLANT AND SOULFUL CHILD I EVER MET.

DAVID M. LEVY, M.D.

*   *   *

Traeger came over a few evenings after the funeral. He had the preliminary autopsy and he told us that slides from Johnny's brain would in time be in every important neurological institute in the world, because it was unique that a child with such an affliction should have remained so comparatively well right to the end. So perhaps Johnny will have aided science after all. Small comfort! It was his courage—the indomitable quality of his simple unquestioning courage—that I hope people will remember him for—that and perhaps for his sweetness too. But what Traeger told us made us glad that he died when he did, so resolutely and peacefully, if he had to die at all, because he would probably have become blind in a short time, and perhaps he would have lost his ability to associate. Even Johnny's gallantry could not have stood up under that. Traeger said, "He had the most brilliant promise of any child I have ever known."

PART TWO

"Recontent with the universe, discontent
with the world."

A Few More of His Letters

———

THE first letter I have in Johnny's own hand was written when he was about seven:

DEAR PAPA

how is your new book getting on, I hope it is getting on all right.

With love and kisses

from

JOHNNY G.

These are excerpts from letters we received while we were away in the Far East in 1937-38:

DEAR MUTTI AND PAPA

I hope you are fine. And so am I. I miss you very, very much. I hung your picture in the bedroom so I can say

[199]

"Good morning" and "Good night" to you. I made a
puppet all by myself it is going to be a woman Detta
is making a dress for her.

We read in the newspaper that a new star was discovered. It is nearly as big as the Solar System.

In school I saw a movie. It was called "If we were
on the moon" It showed that there were many vulcanos
on the moon. One was eighty-five miles wide, that one
is one of the biggest. It showed that you couldn't hear
each other on the moon. It showed that if you weight
one hundred and twenty-five pounds on the earth, you
would weigh twenty-five pounds on the moon.

Our grade went to a Science laboratory. A man
showed us lots of experiments. One of the experiments
showed us that electricity would rather go through two
feet without oxygen than a few inches with oxygen.

This weekend I went to the Van Alens. They had a
big Pirate party. Seventeen boys were invited. A nice
magician showed us some very tricky trickes. I also
went riding on a pony named "Sassy Susie" I had a
wonderful weekend I hardly can wait to hug you

He went to camp in the summer of 1939:

Lake Placid, N.Y.

DEAR MUTTI AND PAPA

How do the basoon double Bass and flute solo records sound.

Ray the science cousler, I and some other children (about 5 I think) banded a Northern Horned lark which is quit rare. He was a baby. the next day we banded 3 baby chipping sparrows.

I am starting to build a modle of the "Soveriegn of the Seas"

I'm fine. Ray caught a porcupine he is keeping him in a big cage there are 50 frogs in the house and at least 11 of them is hopping around lose on the floor. I caught 1 of them. there are four Garter snakes in the house.

I am making a banboo flute. Please send me the violin I have, with strings or no strings. Don't forget the bow. Ray left. I played a littel tennis

Thanks alot for sending the violin I am learning how to play it. the orchestra played its first consert I swam 380 feet on the side strok

[201]

I am having fine time. Thanks a lot for sending the chines things. the camera came in a great big box, there was more wood shavings then camara in it.
I started weaving soming. I moved into a tent on monday one of the workmen caught a woodchuck. he was put in a cage. he escaped today.

Frances had a critically serious illness; I was in Europe covering the outbreak of the war and came home:

DEAR MUTTI

school has been fun. Papa is going to arive on the S.S. amsterdam Friday at 8:oo P.M. Miss chambers is taking Edgar and me to the dock. our teacher is miss Eakright lukaly our project this term is going to be the Geological and sociological formations of USA I am fine. when can you leave the hospital

love

kisses

and

hugs

JOHNNY

Dear Mutti

How are you. I am fine. I am lisening to some good music on the radio by betoven. Yesterday I went to the museum of natural history. and so the hall of gems I so the amsterdam arrive. I hope you will soon be out of the hospitle

love and kisses
Johnny

He wrote me often while I was in Latin America in the winter of 1940-41:

Dear Papa

I miss you very much. Thanks alot for your post-cards and letters I like them alot. Are you still in Panama? I wonder how the lockes work? A week or so a go I heard Hifez he played a modern peice called "the hoxapocha" which is quite unusual I stayed up till about 12 that night I'm having lots fun I hope that you are

love and kisses
Johnny

Dear Papa

We went to Croton for thangsgiving I played a trio

with Whit and peter. Sat. we went to Long Island to visit the Allens Mutti nearly bought a house with ocean to go with it

love

JOHNNY

Frances had a trip to Florida:

DEAR MUTTI

I'm beginning to finish or rather finishing the beginning of that rondo I started compose. I wrote the first theme and a modulation.

When are you coming back? I would like the aligator alot but, I would like to give you the pleasure of letting it accompany you on the trip here.

love

JOHNNY

Then Frances and Johnny spent the summer in California:

DEAR PAPA

I am making a model of a Messershmite 109. What part of your book are you on now.

[204]

I'm learning the second position on my violin. I got a bicycle and I am learning how to ride.

I fixed a contraption where you pull a string in the pantry and a bell rings in my room upstairs.

love
JOHNNY

DEAR PAPA

Aunt Hester drove us to Yosemite National Park. We saw manny water falls Redwood trees and other interesting things. Uncle Bernie had an appendix operation monday. On wednesday he eat roast beef and potatoes.

love
JOHNNY

DEAR PAPA

thanks alot for your letters. I'm getting $.25 a week for polishing mutti shoes

I'm going to the griffith observatory today. They have some rocks with a bulb under each rock. if you think a rock is feldspar you press a button marked feldspar and a bulb lights up under one of the rocks.

Love
JOHNNY

P.S. please send me my chemistry books

This was the first letter we had from him after he entered Riverdale in 1941:

DEAR MUTTI AND PAPA

Since I saw you yesterday, I think it's very silly to have to write you today but since I do I don't no what to write about. I could be doing arithmetic or writing the twopage composition I have to get done by to-morrow.

What about the new apartment?

love
JOHNNY

Here are other bits of letters from Riverdale:

DEAR MOTHER:

I got my copy of "Inside Latin America" and I'am reading it. I just finished the chapter "Hail Columbia" or something like that. I had breakfast with Mr. Hackett again. This afternoon we might go to the radio or we might not. I'll call you up next tuesday, maybe.

Yesterday I paid 50c to see a movie I had seen before (I had to go) and then went to "dots" a drug store

and didn't get anything. I'm planning an electric
motor me and Chris Eaton are going to make.

Last wedensday we went to play another socker
game with another school this time we went to their
field. I'm going to church in a few minutes How are
both books getting on

love
JOHNNY

This came to me in the summer of 1942:

Madison, Friday July 24

DEAR PAPA

Thanks alot for sending the violin A string. I am
haaving a marvelous time out here and I hope you can
join us before going to california Thanks for calling
up.

I am taking both chemistry and tennis lessons I
identified 6 metals or alloys and am making models
of atoms.

Where was your broadcast thursday 7:45? Hope to
see you soon,

love
JOHNNY

By early 1943 he was writing this way:

Riverdale country School

DEAR MUTTI,

I am sorry I didn't come home this weekend but apparently I didn't make the proper arrangements. I will be sure to come next weekend. Shall I reserve Sunday as usual? If I do come over I hope we will finally get that boat.

This weekend I am doing a lot of work on the paper. I have typed two of the stencils myself and Cyril is working on another one. Maybe we can get somebody else to do the others. Don't worry, I am being more careful in typing on those stencils than I am on or your letter. Besides most of the stencil typing is done the other typewriter; this won has its eccentricities. (I hope you don't get jealous!)

Sincerely yours,
JOHNNY

And:

(14 Feb 1943
"Touch" typing—

darling mutti I adore you

These are from letters to me when I was covering the war:

[208]

Aug. 17, 1943

DEAR PAPA,

Happy Birthday! I hope this letter reaches you about August 29, but V-mail may not be as fast from the U.S. as to it. We will send you a cable then anyway, and have a present ready for you when you come back.

The sailboat leaks a bit; the mast has a tendency to bend in a good breeze and the centerboard gets tcusk, but we have a lot of fun in her anyway. Freddy nicknamed the boat "Sinky."

Jerky U Rattler, the 1931 Pontiac, still goes, and Mutti actually went 30 miles per hour in her. My bicycle is in working order and Mutti's is being repaired.

As Mutti wrote you, Chandralehka and Nyantara Pandit were here. Now they are in New York where they are being besieged by missionaries and other people. When they were up here they had to write hundreds of letters which kept them rather occupied.

In the begining of the summer we decided that I should go fishing every Friday and catch some fish for dinner. After a few times I found that I didn't like it because I kept thinking about how I would feel if I were the fish. Mutti thought that I ought to fish anyway and not think about how the fish feels. But when I took her fishing and she tried to dehook the fish, she immediately changed her views on the subject. Now we buy our fish.

We haven't been able to get you on the radio yet but we have seen quite a few articles of yours in the papers. Tonight, however, we will try to get you on the air again.

love and kisses
JOHNNY

These are from Deerfield in 1943:

DEAR MUTTI AND PAPA,

Thanks a lot for those letters and the pen that you sent me. I am having a fine time here doing lots of interesting things—including picking potatoes, I've picked 25 bushels so far. The first Sat. night a magician, Betrand Adams, came here and showed us some fine tricks, but I knew how some of them were done.

Besides the four other subjects I am taking Geology which is very interesting. I found out that I had forgottne most of my latin but I am picking it up again quickly.

All the masters and most of the boys are very nice,

especially Mr. Sulivan. He said that he doesn't believe in punishment, which is always helpful.

Dearest love and kisses
JOHNNY

P.S. Thanks for calling.

DEAR MUTTI

The pears came; and boy do they look good! I can't wait till they get ripe. I have never seen so big pears in my life, and of the 800 known varieties of pears, these are supposed to be the best. Thanks a lot.

Thank you for the clippings you keep sending me—keep it up because every boy has to keep a buletin board filled with clippings, pictures, articles and other things for a week once or twice a term in English History, and the clippings you send me are not only useful for that, but are very interesting for myself.

On Thanks-giving we saw a movie called "Holy Matrimony" which was very funny. I hope you have time to see it. A Mrs. Earl came up to see her son and she told me that she knew me and you and Pa in Vienna. She sends her best regards to you and Papa.

My cold is all gone now and I hope that you are well also and not having any mysterious ailments.

love and kisses,
JOHNNY

1944 opened with:

Jan. 10

DEAR MUTTI,

Here I am back at the old grind again! But, as always, it is much better than I thought it would be. This term is only eight weeks—and then comes three more weeks of vacation.

I can not find any snow-shoes anywhere in my room, and I am positive that we didn't bring them up. There is plenty of snow, however.

Love,
JOHNNY

Jan. 30

DEAR PAPA,

Believe it or not, I don't believe I forgot to take along a single thing; but the infirmary would like to know the results of the blood test I had at Dr. Traigor's, so would you ask him to send them up.

Thanks ever so much for the wonderful time I had during the vacation—I don't really think that I ever had more fun on one. I'm learning to do parralel christies. Also, I'm trying to improve my grades, which Mutti said could stand improvement.

dearest love
JOHNNY

The following are bits from letters to Frances:

DEAR MUTTI,

I read "Wind, San, and Stars" again and I think I understood it much better this time—and I really loved it. I gave an oral book report on it and got C+/A (C+ for the presentation and A for what I said). Alltogether I'm doing much better in English so far this term. On one thing I even got an A in the top (grammar, spelling, punctuation etc.) mark. (You know how we get 2 marks in everything we do in English.)

I guess you will have gotten my marks by the time this letter reaches you. Remember that both of the English marks together only count as much as one of the other three. I got 97% on last term's Physics test, and may even get A for the term.

I hope your room isn't too "painfully" neat. If it is, you know what can be done about it! Of course my room isn't "painfully" neat, but it isn't at all on the other extreme. In fact Mr. McGlynn, our corridor master, even commented to that effect, and if you were to see my room as I am writing this letter you would be quite astonished, for there isn't anything extraneous either on my bed or on the floor.

Did you notice in the day before yesterday's (Thursdays) Times that the British Battleship Hood was not

sunk by the Bismark, but by the cruiser Prinz Eugen?
I've enclosed the clipping.

love

JOHNNY

P.S. I sent that question to "Information Please"

I've been doing my eye-exercises almost every day,
washing my face diligently, and putting that sulfa oint-
ment that Dr. Traeger prescribed on my spots and
blackheads; and I try to be spontaneous, uninhibited,
self-aware, and self-controlled;—but those vitamin
pills, those sublimely wonderful vitamin pills!—well,
thanks, for reminding me of them anyway. At least
I hadn't lost them, and I've really taken one every day
(well, every day but one or two) since I got your letter.

I've been doing quite a bit of work on my stamps
recently. I bought a bunch of Indian stamps, most of
them very old, for $12 "blind," that is, without look-
ing them up in the stamp catalogue, which would have
taken several hours, so that neither I nor the fellow I
was buying them from knew how much they were
worth. They turned out to be worth over $20! Don't
worry, the guy has gypped me several times too. I think
I must have quite a valuable Indian collection now,
including all the airmail stamps that India ever
issued, and two of the only four stamps which the state

of Alwar ever issued, and a stamp which would have been worth $75 if it wasn't fake. Don't worry, I didn't pay $75 for it!—it came in that group which I bought for $12.00.

I've played a great deal of chess recently, and am No. 2 on the chess team. We had a match with the Greenfield Chess Club, most of whom seemed to be octogenarians at least, and they trounced us something like 8½–3½. I was lucky to get a draw. I hope you have looked at that chess book I gave you, for it's a very good one.

Thanks very much for that marvelous candy and that white shirt. That laundry hasn't come yet but I have written to Nancy about it. There couldn't have been much in it anyway because I seem to have plenty up here. The laundry often sends me a few of somebody else's shirts anyway!

Has all the fertilizer arrived? And what about the water pump and the pruning and the spraying of the trees? I haven't written Mr. Weaver about his fertilizer yet but I hope to soon. I might be able to take a weekend to do some more work on the place and perhaps plant some seeds, but I think that had better wait until after Spring Day (May 12).

As for tennis, I'm afraid it's entirely competitive,

and about a third of the boys going out for it were eliminated. Apparently they couldn't get equipment for the new courts. I am going out for lacrosse instead and I'm really not doing so badly in it and it's lots of fun. I couldn't quite face track. . . . We have a test in Biology every Friday, and I've got 100% in all three so far.

I try to remember those little lectures that you gave me during vacation. Don't get discouraged about them because I usually remember the general idea, sub-consciously if not consciously. Sometimes after coming to a certain conclusion for a course of action after a long train of thought I think Good Lord, that's what you've been trying to tell me for the last year.
P.S. I'm breathing out copiously too.

With love and spontaneity
JOHNNY

These are from notes to me:

DEAR PAPA,
A certain Mr. Mayer, father of one of the boys here, who was in Switzerland at the American Legation, came up here and told me that he knew Aunt Billi, who gave the message that she was well and happy,

and said that he could bring back a letter to her. He is leaving soon so if you write a letter to her and send it up here maybe it can be sent to him.

Thanks for that letter and the clipping. Is Texas really that tremendous? (or do you say that about every state?)

I arrived safely, and so, in fact, has everything else— the things from Best's, the trunk and even the Smith report, which is fascinating, but I haven't had a chance to read it carefully yet. The lamp is not flourescent, but PERFECTLY adequate. It contains a 60 watt bulb, a shade, and some very flexible metal tubing (like the old broken one) so that I can shine it on any part of the desk I want to, even the typewritter, which is more than I could do with a flourescent one. Tell Mutti that people read and wrote before there were flourescent lamps.

I have graduated from waiting on to dishwashing (usually referred to as "slops") which is much better— the total number of meals to do is less, and it's not nearly so nerve-racking. You don't actually wash the dishes; they have to be sent thru a machine which theoretically does it for you. It really does fairly well, unless there is something like peanut butter. But we don't have peanut butter very often anyway.

I have also graduated, at least temporarily, to the Junior soccer squad. For the first time I am not on the lowest team of any particular sport since Lincoln. I hope I stay there. Our coach was also my English teacher my Freshman year and he is very nice.

I have an ideal schedule, with study halls and subjects alternating all the way thru. The Physics teacher gayly told us it would be very easy if we spent two hours a night on it, but for me anyway it hasn't been at all difficult so far. In French we have a teacher who's been to France and really knows the language.

I have plenty of clothes so never mind the rest of the laundry. Good luck on the last 8½ states!

Thanks for all three of your letters with those clippings. I did read that story by Wylie and last summer I read "A Generation of Vipers" by him also. Quite a book! But Mutti told me not to take it too seriously and I didn't, so don't worry over any affects it may have had on me.

The following are excerpts from letters to his mother in the autumn of 1945:

DEAR MUTTI,

Papa got the Smyth Report on atomic power for me and I am reading it. I will explain it all to you when vacation comes again. Do you know that no actual particles of matter (electrons, protons etc.) are turned into energy in the reaction? Every element weighs slightly less than the sum of the weights of the neutrons, protons, and electrons of which it consists. This loss in weight is known as the "binding energy" and it is less in U235 than in the elements into which it breaks up, so the total mass is lowered, and a corresponding $(E > mc^2)$ amount of energy is released.

I've joined the Chess Club, which has started already, and have played a number of games. Aside from that, I won a dollar by beating Steve six games straight. I don't as a rule play chess to enhance my financial assets, but I offered the bet (even money that I could beat him six straight) as a sort of joke and didn't want to back down when he accepted.

I got 95 and 98 respectively on the only big physics tests we've had so far—I hope I can keep it up. I've also been reading a great deal in the Science Library on Physical chemistry, atomic physics, the quantum theory, relativity and sundry other subjects like that which seem to have a peculiar fascination for me.

Every now and then I get grandiose ideas for experiments or theorys which would astound the world, revolutionize science, etc., etc. Also, on several occasions I have suspected things which later, by deeper delving into the subject, I found to be true; something which is very gratifying.

The package to which you referred came the next day. What a beautiful bathrobe! Thank you very much.

These are the last two letters I had before his illness:

Sunday, Jan. 13

DEAR PAPA,

I arrived safely and have already started counting the days to the Spring Vacation. The winter term is the shortest one and we only have about forty actaul school days. The suitcase has arrived, but the coat hasn't, but I guess it will in a few days.

Thanks a lot for taking care of the experiment. As long as there was any water in the acid the reaction probably went like this:

$$2H_2SO_4 \rightarrow 2H_2\uparrow + 2SO_4 \searrow$$
$$2SO_4 + 2H_2O \rightarrow 2H_2SO_4 + 0\uparrow$$

[220]

The net result being the elimination of water as hydrogen and oxygen. After all the water was eliminated the action either stopped there or continued thus

$$2H_2SO_4 \rightarrow H_2\uparrow + H_2S_2O_8$$

(Sulphuric   (Persulphuric
acid)   acid)

the net result being the changing of sulphuric acid into persulphuric acid and hydrogen. When all the sulphuric acid was eliminated it may have stopped there or God knows what may have happened before it finally stopped. Anyway, I'm awfully glad you took care of it so well.

love

JOHNNY

P.S. Coat just arrived—marks will go out in a few days

April 16

DEAR PAPA,

School again!—but its really not so bad after all; in fact, its lots of fun, no matter wtat you think over vacation. I found my key, just where I thought it was, and have resolved not to forget it again. Thanks for that clipping on Alekhine.

I got 94% on last term's final physics test, B for the

[221]

term in French, but I'm not sure of my other term marks.

My prize for winning the "Time" Current Affairs Test came a few days ago. I was lucky to get my first choice, "An Outline of Atomic Physics." Of course I would have to get a book that was meant for those who had taken a year of college physics. It takes about half an hour to make out each page, but I'm struggling with it diligently. As for athletics, I'm going out for soccer again, as the least of five or six evils. Luckily, the coaching is much better than it was in the fall.

I suppose you saw in the papers the other day that one Mr. Olsen was preparing to run for the Democratic nomination for Guvernor in Nebraska after (and this you may not have noticed) having worked in the kitchen of Deerfield Academy for several months. I supose he is of no particular importance, for he certainly doesn't seem to have any political opinions whatsoever, but he is a jolly old man with a wonderful gift for making himself popular and well-liked.

dearest love,

JOHNNY

·

This was written to a schoolmate two weeks after the first operation:

May 14, 1946

DEAR STEVE,

Thanks for that letter. I should like very much to clear up any misapprehensions you and the Corridor may have about my state of affairs. Frankly, I think I have discovered Utopia here at the Presbyterian Hopsital. No school work. No athletics. No worries. All I do is eat, sleep, and have a wonderful time generally. My parents visit me every day. (I am dictating this to my mother.) My reading is still somewhat restricted, but soon I shall be able to have any book I want.

I had quite a serious operation. They had to drill three holes right through my skull. I'll bet you'd never guess what the trouble was—excess pressure within the brain. The most painful part of the operation occurred beforehand—they had to shave off all my hair! That hurt so much that I resolved to grow a Brahmsian beard and never shave in my life. My hair is growing back—slowly.

Give my best to Mr. McGlyn and all the boys. Show them this letter—unless you think they'll get too jealous.

This was in longhand to another schoolmate:

DEAR JOHN, Thanks for your card—you say you went to a camp. I must say [*undecipherable*] seems to me to be peculiarly machiavelian form of torture [*undecipherable*] able to give and they gave me 50 days worth of X-rays in 18—I always say I've got a wooden head and an iron stomach. Since then I had a wonderful time here in Madison though I couldn't swim or sail or do anything very strenuous. Except about two weeks back at the hospital. They gave me a very new treatment—a derivative of mustard gas and I made a very good guinea pig. It was the first time it was tried at Presbyterian hospital, though several hundred cases were tried at other hospitals. I only puked three times—something of a

Then:

July or August '46

DEAR MR. BOYDEN,

Thank you very much for your letter, and particularly for that picture of Dr. Einstein. As for that crazy letter I sent to him, I only wish I could understand it myself. As I wrote to Mr. Haynes, I think I must have written it under the influence of an overdose of caffeine, which I took to relieve headaches.

I will write Mr. Miller about my room next year as you sugested.

Would the following set of courses be satisfactory for me next year?

> English IV
> Chemistry
> American History
> Mrs. Boyden's Math Gamma

I spent a wonderful month in the country, but now I am back in the hospital, but only for a few days (we hope) I've been doing considerable review work in math and physics.

DEAR MR. WEAVER,

Thank you for your kind letter—which my Father and Mother also enjoyed and appreciated very much. Here's hoping your boys are studying hard and working well and not blowing up the place!

Give my best to Andy and Mrs. Weaver and the children.

I have been interred for weeks now subsisting on a diet of abominations. All I get is squash, brussels sprouts, turnips, spinach and the like—and I get no proteins, no fats, no salt—they give me dozens of injections and myriads of pills—(If you ask me, they are all laxatives too!)—and tons of fluids—fruit and veg.

juices. However, it really isn't so bad, and I suppose I shouldn't complain.

They tell me it is doing some good and I hope to be back to school soon.

You may be interested to know that I was very lucky to meet Mr. & Mrs. Lieber who have written the books on higher Math in verse all full of wonderful illustrations. Mrs. L. is head Math Dept. and Mr. L. of Art at Long Island University. And Dr. Francis Bitter dropped in from M.I.T. and we had a fine talk with him.

Give my best

Nov. 18, 1946

DEAR STEVE,

Thanks for your letters, and I trust you had great fun picking potatoes! As for me, I have to eat the darn things (dry and unsalted), and don't even get the satisfaction of throwing them. I am completely overwhelmed with myriads of pills—plus the stuff they call food here (I am still at the special clinic just for this diet). But I am much better and will be able to move home soon.

I've turned over a new leaf! I've forgotten all the theories that I tried to inflict on you last year and I've abandoned the pseudo-scientific brainstorms that wasted all my time last year. I'll have my hands full just with schoolwork. I'm going to try to take my intermediate algebra final exam, so keep your fingers crossed for me.

Give my best to the boys.

Sincerely yours,

JOHN

Nov. 18, 1946

DEAR MR. WEAVER,

Thanks so much for finding out about my experiment, and also for your fine book and letter.

Now that I've gotten used to this diet, I find that it really isn't so bad after all; and after 28 months I will be able to have an egg once a week!

I've mastered the binomial theorem and after some more review work will try to pass my intermediate algebra exam.

Best wishes to Mrs. Weaver and the boys.

Dec. 1, 1946

DEAR MR. BRIDGMAN,

Would you be so kind as to inform me what English teacher I will have this year. I am feeling much better now and I think I will be back at school for the winter term.

Yours sincerely,
JOHN GUNTHER, JR.

Dec. 11, 1946

DEAR MR. WEAVER,

A million thanks for your letter! Next time you see Dr. Foster please thank him for me. I will write to the man you suggested. I do hope all this hasn't caused too much trouble. I've written up twenty-five chemical experiments that I've done in the last five or six years, including many that we did together. (Remember that dust explosion?) I hope this will take care of last term's chem. lab.

I'm back at Neurological Institute now after having been at a special clinic for my diet. All the doctors here are very impressed with my improvement.

Best wishes to Mrs. Weaver and the boys.

Sincerely yours,
JOHNNY

During 1947 Johnny wrote little. One of the last scribbles I have of his handwriting is a note: "Scientists will save us all." His last letter to his mother is in the text above.

[228]

The Diary

———

SPORADICALLY for some years Johnny kept a diary, at Frances's suggestion and under her encouragement. When he became ill he carried his notebook wherever he went; on any trip to the hospital, no matter how brief, two things were musts: the recorder and his diary. He told her once, "You put it on my desk so gently. You didn't tell me to use it. You just put it gently on my desk, remember, and then I began using it, and I'm so glad."

Usually he wrote secretly. But then he would push the notebook in a drawer, leave it unlocked, and make casual references to it. We think now that perhaps he was using the diary, such as it was in the later stages when his handwriting had become pitiably marred, as an indirect way of conveying messages to us. He was inviting us to read it and so find out what he was thinking that he didn't wish to talk about.

But first here are some early notes, written long before his illness:

October 19, 1944
Woke up. No dream

Tuesday March 20th, 1945
Worked on garden. Saw Mr. Weaver's cousin and good watch dog. Bicycle needs repair. "A Tree Grows in Brooklyn." Cheerful.

Thursday March 22
Arthur Murray—Miss Cummins. Not so bad after all. Have to face it sometime.
List of magic tricks as perfected:
Red and black
Coincidence
Favorite number
7 and 8
Chinese rice bowls
Sugar and water
anti-gravitation

Friday March 23
Ammonium sulphate came. Finished little plots. Fertilizer arranged. Saw "Carousel."

There follows an item that he got 100 in a geography quiz and scored his first goal at lacrosse.

Tuesday—April 3

2 tricks learned:

    Card on ceiling

    Slap card out of hand

Still nervous, tense. Cannot be myself in critical moments. Happy.

The adjective I like least for myself is naive.

Thursday (otherwise undated)

A.R.I. Wylie's article in R.D.—never decline challenge

All of the following notes are from the summer of 1945:

Friday

Started repairing wheelbarrow. Did considerable work on rocket plane. Wrote letter to grandma.

Life is short—so waste not a minute.

Tuesday

In New York. Eye exercises. Started clean apt. of chemicals.

Monday

Steve: Sue & John Hubner here. Tested Beale's soil. Saw Mutti's stories in News-Chronicle. Arranged monthly system of French word lists.

Wednesday

Went out all day on sailboat. Couldn't get to Faulkner's—had picnic on Tuxers. Mutti shocked when she learned we tried for Faulkner's—Coast Guard, police, etc. called. Misunderstanding at beginning. But I lacked gumption—then refused permission to retry for Faulkner's. All ended happily.

Gunther Philosophy:

1. Do unto others as you would have them do unto you but don't love God with all your heart, mind, sole. To really believe in first you don't need second. Good is an axiom.

2. No immortality

3. Live while you're living then die and be done with—never refuse challenge—never give up trying etc.

4. "How to Win Friends and Influence People" idea (Say nice things to Mutti once in a while).

5. Happiness depends on moderation—not too much ambition — Humility — "don't-give-a-damn-ness"

Saturday

Fixed wheel cart. Worked a lot on grass. Read "Lost Horizon" and Stefandson's article on Eskimos. British elections—!!

Sunday

Hubners and Helena Scheu here—beat latter in chess—learned a lot.

Wednesday

Put up sign but didn't paint over other one. A lot of work on garden. Game of poker. Experiment— Putting magnet in colloidal graphite and salt solution. Wrote Pop.

Saturday

Long sail with Johnny. Taught him chemistry— electron theory etc. List metals & periodic table. Slept —ate—various errands.

Sunday

Before every act—ask yourself if that is what you want to do.

Tuesday

Good day. Lots done. Workshop.

Wednesday

Self analysis: About $\frac{1}{2}$ time my conscious mind is either asleep or wandering off in space. This . . . accounts for procrastination etc. I am greatly over-introvert—caused by over-consciousness of what others think of me . Caused by my atheism (?).

Friday
V-J Day

Sunday
Sailed all day long.

Tuesday

War ended.

This is the first entry after the illness, written five days after Putnam's operation:

May 5, 1946

More wonderful experiments! Compare M-N **and** shifting of red line. See Russell on de Sitter.

May 26, 1946. 9 A.M.

Have to stay in bed until temp. goes down. One minute later. Sitting up in bed!!

Last night—idea for space and time-like intervals came to me.

Discovered physiology of cough, how to cough silently—slow motion—what fun with the doctors.

Does phosphene react with anhydrous sulphuric acid in a manner analagous to the way ammonia reacts with water?

Interlude. Experiment. Ph. Analysis H_2SO_4.

By various indicators electrolysis to dehydrate H_2SO_4. Solubility KCL in H_2SO_4. Solubility $AGNO_3$ in H_2SO_4.

May 30, 1946

Hurrah. Finally saw how [two words indecipherable] arose from concept of proper time.

Friday, 31st

I guess Dr. Einstein saw through my letter, but it really was a pretty good job of bolstering me. How kind a man he must be to bother to reply anyway. Proton: a "hollow" in space time, where the number of coordinates necessary becomes infinite.

Summer, 1946

Wonderful time for a month or two in the country. Otherwise in the hospital—first at Medical Center, then at Mrs. Seeley's clinic. Fall—at clinic taking this diet. [Then comes a section marked "Interlude from last summer" with a long list of chemicals and apparatus and drawings of various experiments to be set up.]

November 10, 1946

Seventeen years old now. Finally "accepted" Dr. Gerson's diet after two months.

November 11th

Ask parents what you can do to make them happy. Mother and Father (that's how I call them now, not Mutti and Papa).

Dr. Gerson says reducing diet overdose may have caused liver trouble and that in turn the tumor. This troubles me. Recently turned over a new leaf. No more brainstorms. From now on I do school work.

Mother and Father swapped apartments a few days

ago, preparatory to my moving from here to 530. Remarkable smoothness.

Always:

Work, talk, give.

November 12, 1946

Talk. Give. Work.

Here is a prayer I though of last spring at Medical Center.

Live while you live, then die and be done with.

November 13th

Liebers were here. Mr. Lieber showed me psycho-scapes—spontaneous drawings with crayon.

November 16th

Resolved to ask Father about divorce.

A few days ago Mr. Haynes came down from school to see me. I had written a letter of despair to Mr. Boyden. Mr. Haynes reassured me. Let me off some of my lab work.

November 17th

Got Father's and Mother's sides of divorce all straightened out. What wonderful parents.

November 18th

Wrote three letters, to Steve, Mr. Boyden, Mr. Weaver.

Selves:
1. Ambitious—sanguine (not very sensitive)
2. Walking-up-and-down-self
   Sensitive, clever
3. Meditative

November 19

Mr. Ohl, my tutor, came for the first time to the clinic. It is a private home which Dr. Gerson uses as a clinic. Mrs. Seeley, Miss Gerson—such nice people!

November 22nd, 1946

Wrote letter to Mr. Haynes.

Philosophy: "Get yourself off your Hands." Happiness is in Love. Accept disappointments. Relieve oneself by confession of sins. I am growing up at last.

November 23rd

Took trial examination in intermediate algebra by myself, on my own. I am trying to keep up with this year's English and history while tackling last year's exams one at a time.

I must ask Mother what I am still "failing" in, such as posture, etc. Dr. Traeger visited.

Things to do:

Make list of things wanted in the country.
Spelling rules for Edgar.
Lab work.

November 29th

Henry, John D. and Judy were here.

I have yet to learn the lesson of contentment.

I think I forgot before, I had a wonderful birthday party, even ice cream, though not very good.

Accept misfortune.

Get yourself off your hands!

Be spontaneous and aware. LOVE!

November 30th

Steve Stullman here from school—Thanksgiving vacation.

I guess I'm not the dreamy neurotic old fool I sometimes think I am.

I thought of 47 chem. experiments which I've done in the past year or so. It will take care of my chem. lab for two terms. Hurrah!!

I will write them up soon.

December 1st

I am rather tired today—didn't do much of anything. I shave every two weeks or so. I try to be "aware" of posture, pills, telephone conversation, facts and figures people tell me.

December 4th

Back at Medical Center.

December 6th

My nurse, who I only know two or three days, compliments me quite overwhelmingly on my personality.

This cheers me up. Keep up the good work. People really appreciate it.

December 10th

In a few hours I will try last year's algebra exam. Get news about experiment NaH elec. in NH_3.

December 11th

Dr. Gerson, Mount, Miller all here. Changed bandage. I did very badly yesterday on exam. Wrote letter to Mr. Weaver. Mrs. Seeley visited me.

December 12th

I am "aware" of spelling, vocab, now. Posture improved. Mother has been showing Billie and Sue how to bring food from the Seeleys for me.

Friday, Dec. 13th

Book-of-the-Month Club accepts Father's Book!!!!! No more financial worries!

December 14th

Mother has to bring food every day from Mrs. Seeley's now. Then she stays the afternoon. Father comes up in the evening.

Today I was too lazy to get dressed. Tomorrow I will.

December 16th

Still at Neurological Institute.

Billie brought up the food for Mother today. My

lab work chem. (this year's) and physics (last year's) is finished and I showed it to Mr. Ohl. Now I will do a little algebra in case I flunked the exam, which Father sent to school.

I asked Father to call up Dr. Gerson's assistant to see if NaCl irrigation is all right. Perhaps they can use KCl instead. Gerson gets apoplexy at the sight of sodium!

December 17th

Ask Francis Bitter what is the relation of that which a particle loses upon deceleration and that which it gains upon the approach of another particle.

I have been much more "spontaneous" ever since the operation last spring. I relieve myself of worries regularly to Mother.

Now I can't even remember the things that used to bother me.

This annoys me.

December 18th

John Davison here. He told me about Mrs. Boyden's special class, Math Beta—I must find out more about it.

December 21st

Hubners visit me. I do a lot of physics and math. Father can't come. Mother tells me of Grandma. This tumor is wearing her down somewhat.

I asked Dr. Mount the other day to dig another hole in my head and pull the packing through. I have an abscess from the dead tumor cells. I have had surgery, x-rays, mustard gas and Gerson diet.

December 22nd

I teach Edgar chemistry lessons. Henry visited me. Tomorrow Mr. Ohl (my tutor) comes. I did lots of work.

My papilledema is gone, but my visual field is still restricted. They are finally going to stop the NaCl in my irrigation. I called up Dr. Traeger about it.

Every day I call Mother and Father in the morning. I get dressed and bathe almost every day now. I drink one and a half quarts apple-carrot juice, one quart soup and vegetables, unsalted. No fats, little protein.

December 23rd, 1946

Dr. Mount gave me permission to go home for 24 hours!!

I must get: Physics test. Dry shampoo. Clothes. Notebook.

I shaved today. Zurs are coming. Write Xmas cards.

December 24, 194

I go home for 24 hours. Xmas Eve I play Christmas carrols on my recorder.

I must write an essay on education for Atlantic Monthly contest, also on "Bull Slinging," Lincoln's defect, and Silas Marner.

December 24–26

Judy, Hubners, Frankie Adams, Francis Bitter, David Burnett, are all here.

Wonderful presents!

    Two Thesauruses

    Watch

    Dictionary

    Wonderful pen

I help with tree and all—and write thank you notes! —also get stuff from Mrs. Seeley.

Had wonderful lunch there.

December 27th

Dr. Mount changes my bandage. I have a big one covering my ears but nice and loose. My head itches!

I call up Mr. Boyden!—find out about Math exam. I filled out my address book.

December 30th

I passed the math exam! Now for physics.

Francis Bitter is here. He explained a scientific toy that has been driving us insane. A little glass duck that moves by itself perched on a glass of water. I worked on physics too.

I remember "Wind Sand and Stars," The Yearling," etc. Have I read "Black Boy"? I can't remember. I guess I'm a bit tired and given to introspection today.

I am at last learning the art of conversation. (I hope!) Up until last spring I hardly talked at all. Mother used to get awfully mad.

Before going to bed I call up my parents.
I feel strangely content now.

Facts about Operation

Face swollen so I couldn't see four to five days.
"Needle in arm" three full days without my knowing it.
O_2 in my room for four days; used it for 36 hours, it cost $100.
When my father told me I had had a little operation, I said, "Of course, I heard them drill three little holes."
Before the operation I had a Wassermann!!
First thing I asked for is my physics book.

January 1, 1947

Yesterday I cleared up the whole matter of the Jews with parents.
Today John Davison and Edgar were here.

January 3rd

Passed math exam. Billie here today. Worked on physics.

January 4th

Called up Cass Canfield to thank him for present.

January 5th
EFFICIENCY!

[243]

January 6th

Had a talk with Professor Shaub, fellow patient. I must tell him my theories, etc. Korzybski is too much for him also!

Dr. Gerson here. I still have the little heat machine that was given to me (a duck-shaped scientific toy). I showed him how it works—it drives everyone crazy.

January 8th

Yesterday I discussed fears of death with Mother. For years I have had a lack of confidence in myself, fears about ultimate reality.

Accept death with detachment.

Take more pleasure in life for its own sake.

PROFESSOR SHAUB TEACHES ME CALCULUS!

MY QUESTION USED ON INFORMATION PLEASE!

January 16th

I am back at Mrs. Seeley's now. Have been for three–four days.

Recontent with the universe. Discontent with the world.

January 17th

Letter to Aunt Hester. Edgar visited me. He complains of his school work.

December January 18th

Finish review of last year's physics except for last-minute review. Day after tomorrow I go up to Medical Center for change of bandage. We had a little party here.

Jan 19
Tired and worried until Mother straightens things out.

January 21st
Dream—no room for me when I get back to school.

January 24th
Card comes from Henry in our chess game by mail. Bertha corrects chemical experiment; I can spell after all! Once I had a bad English teacher. . . . therefore I didn't do any work in it. Got terrible marks.

January 26th
Yesterday I took last year's physics exam.

February 3rd
Sometimes I wish I was as cheerful to myself as to others——nonsense!!

Oh! What a letter I wrote to "the Bart" [Mr. Boyden]—500 words vindicating myself.

We have an English tutor for me now, Mr. Seton. I work very hard.

Dr. Gerson has great fun with my metabolism. It's too high, then too low, then too normal.

[245]

February 9th, Sunday

HOME!—since last Thursday.
Dinner Party!! Saturday
MARY SANDERS HERE!
(My temp. goes up to 99.5!)
And Edgar, Judy.

February 17th

Oh how I work on last year's English! Mr. Seton tells me to slow down. We go over to Mrs. Seeley's for lunch almost every day.
Metabolism is right!—plus 13.
I give myself injection!
Worried about dancing.
Encyclopedia arrives!

Wednesday, Feb. 19, 1947, A.D.

A little amnesia today.
I think that I realize and accept the "goodness of life." I should not need to "hang on to" chemical brainstorms, self-abnegation, etc.

Tuesday, March 18

The other day I saw Dr. Levy again. I am up to Civil War in the history.
Things to do; autobiography. Scroll. Call up Judy and Lionel. Clean up. Henry comes Thursday.
Dr. Levy—Friday 11. Dr. Berliner—Tuesday 2:30.

April 13th

Yesterday I took seven hours of college board exams! The other day Father took me to the Public Library to look up facts about lithium hydride and liquid NH_3.

I am still not permitted to go outside by myself.

Monday 21

Oh how tired I feel.

Bulkley over here. Tells of his escapade.

May

Back to Neurological Institute! A second operation. They shave off all my hair again! Damn it.

But I can eat again! Steak, ice cream! Cream of mushroom soup!

Oh! How good it is.

May 13th

Yesterday I got outside for a walk with Jim Cortwright and nurse. Also I worked hard on trig and English.

I get letter from Mother in Florida.

May 16

Home again after only two weeks! I go to tutoring school.

In hospital after brain operation the other day I played poker with fellow patients.

May 20

Parents go on successive vacations. I work hard on trig. A letter comes from Bob Harrison.

May 30th

Parents drive me back to Deerfield! Oh! It's great! I am given a German book and a math book. Mrs. Boyden tutors me in chem. Mr. Havilland in German. I roam about.

My left hand is still a bit clumsy. I stay at the Infirmary nights. I can't yet tie my shoe. I brought the metallic lithium up with me.

I say hellow to all the boys.

June 3rd

We (Mother and I) see The Mikado, the Senior play.

Heywood Alexander is wonderful as Kiko.

I graduate—get diploma!

June 5

Home again, 530 Park!! I look up things in the Public Library about liquid NH_3. I lunch with Jon Van Winkle, a new boy whom I met in my week at Deerfield. He had taught me some calculus.

June 12

I go back to——Memorial Hospital for nitrogen mustard treatment.

[248]

I finish reading "Human Destiny" by Du Nouy. It proves existence of God, or an "anti-chance," from evolution and by scientific reasoning.

Thursday, June 26
I go to Public Library to collect information on liquid NH_3.

Yesterday I saw Peter Blose after all these years! He and Edgar and I romped together in Fourth Grade at Lincoln.

I phone Mary but she can't come to lunch.

These are the last entries in the diary proper, four days before he died.

At the back of the notebook are several pages of notations of chemical and mathematical formulae, also a list of presents he wished to give Frances and me. I was to get a pipe, some records, and a size $7\frac{3}{8}$ hat; Frances was to have an electric blanket, a large candle from Georg Jensen, and an old Fifth Avenue bus.

The very last words are written inside the back cover. Frances had told him the story of the ancient Hebrew toast, "L'chaim." Johnny's notation is: "Hebrew Toast: Le Hy-eem—To Life."

A Word From Frances

———

DEATH always brings one suddenly face to face with life. Nothing, not even the birth of one's child, brings one so close to life as his death.

Johnny lay dying of a brain tumor for fifteen months. He was in his seventeenth year. I never kissed him good night without wondering whether I should see him alive in the morning. I greeted him each morning as though he were newly born to me, a re-gift of God. Each day he lived was a blessed day of grace.

The impending death of one's child raises many questions in one's mind and heart and soul. It raises all the infinite questions, each answer ending in another question. What is the meaning of life? What are the relations between things: life and death? the individual and the family? the family and society? marriage and divorce? the individual and the state? medicine and research? science and politics and religion? man, men, and God?

[250]

All these questions came up in one way or another, and Johnny and I talked about them, in one way or another, as he was dying for fifteen months. He wasn't just dying, of course. He was living and dying and being reborn all at the same time each day. How we loved each day. "It's been another wonderful day, Mother!" he'd say, as I knelt to kiss him good night.

There are many complex and erudite answers to all these questions, which men have thought about for many thousands of years, and about which they have written many thousands of books.

Yet at the end of them all, when one has put away all the books, and all the words, when one is alone with oneself, when one is alone with God, what is left in one's heart? Just this:

I wish we had loved Johnny more.

Since Johnny's death, we have received many letters from many kind friends from all parts of the world, each expressing his condolence in his own way. But through most of them has run a single theme: sympathy with us in facing a mysterious stroke of God's will that seemed inexplicable, unjustifiable and yet, being God's will, must also be part of some great plan beyond our mortal ken, perhaps sparing him or us greater pain or loss.

Actually, in the experience of losing one's child in death, I have found that other factors were involved.

I did not for one thing feel that God had personally singled out either him or us for any special act, either of animosity or generosity. In a way I did not feel that God was personally involved at all. I have all my life had a spontaneous, instinctive sense of the reality of God, in faith, beyond ordinary belief. I have always prayed to God and talked things over with Him, in church and out of church, when perplexed, or very sad, or also very happy. During Johnny's long illness, I prayed continually to God, naturally. God was always there. He sat beside us during the doctors' consultations, as we waited the long vigils outside the operating room, as we rejoiced in the miracle of a brief recovery, as we agonized when hope ebbed away, and the doctors confessed there was no longer anything they could do. They were helpless, and we were helpless, and in His way, God, standing by us in our hour of need, God in His infinite wisdom and mercy and loving kindness, God in all His omnipotence, was helpless too.

Life is a myriad series of mutations, chemical, physical, spiritual. The same infinitely intricate, yet profoundly simple, law of life that produced Johnny—his rare and precious soul, his sweetness, his gaiety, his gallantry, his courage: for it was only after his death, from his brief simple diaries, written as directly as he wrote out his beloved chemical experiments, that we

learned he had known all along how grave was his illness, and that even as we had gaily pretended with him
that all was well and he was completely recovering, he
was pretending with us, and bearing our burden with
the spirit, the élan, of a singing soldier or a laughing
saint—that law of life which out of infinite mutation
had produced Johnny, that law still mutating, destroyed him. God Himself, no less than us, is part of
that law. Johnny was an extraordinarily lovable and
alive human being. There seemed to be no evil, only
an illuminating good, in him. Everybody who knew
him, his friends and teachers at Lincoln, Riverdale,
and Deerfield, our neighbors in the country at Madison,
felt the warmth of his goodness and its great vitality in
him. Yet a single cell, mutating experimentally, killed
him. But the law of mutation, in its various forms, is
the law of the universe. It is impersonal, inevitable.
Grief cannot be concerned with it. At least, mine
could not.

My grief, I find, is not desolation or rebellion at
universal law or deity. I find grief to be much simpler
and sadder. Contemplating the Eternal Deity and His
Universal Laws leaves me grave but dry-eyed. But a
sunny fast wind along the Sound, good sailing weather,
a new light boat, will shake me to tears: how Johnny
would have loved this boat, this wind, this sunny
day!

[253]

All the things he loved tear at my heart because he is no longer here on earth to enjoy them. All the things he loved! An open fire with a broiling steak, a pancake tossed in the air, fresh nectarines, black-red cherries—the science columns in the papers and magazines, the fascinating new technical developments—the Berkshire music festival coming in over the air, as we lay in the moonlight on our wide open beach, listening—how he loved all these! For like many children of our contemporary renaissance, he was many-sided, with many loves. Chemistry and math were his particular passion, but as a younger child at school, he had painted gay spirited pictures of sailing boats and circuses, had sculpted some lovable figures, two bears dancing, a cellist playing, and had played some musical instruments himself, piano, violin, and his beloved recorder. He collected stamps, of course, and also rocks; he really loved and knew his rocks, classified them, also cut and polished them in his workshop, and dug lovely bits of garnet from the Connecticut hillsides.

But the thing closest to his heart was his Chem Lab which he cherished passionately. It grew and expanded in town and country. He wanted to try experiments that had not been done before. He liked to consider abstract principles of the sciences, searched intuitively for unifying theories.

He had many worthy ambitions which he did not live long enough to achieve. But he did achieve one: graduation with his class at Deerfield. Despite the long illness that kept him out of school a year and a half, he insisted on being tutored in the hospital and at home, taking his class exams, and the college board exams for Harvard, and then returning to Deerfield for commencement week. The boys cheered him as he walked down the aisle to receive his diploma, his head bandaged but held high, his young face pale, his dark blue eyes shining with the joy of achievement. A fortnight later, he died.

What is the grief that tears me now?

No fear of death or any hereafter. During our last summer at Madison, I would write in my diary when I couldn't sleep. "Look Death in the face. To look Death in the face, and not be afraid. To be friendly to Death as to Life. Death as a part of Life, like Birth. Not the final part. I have no sense of finality about Death. Only the final scene in a single act of a play that goes on forever. Look Death in the face: it's a friendly face, a kindly face, sad, reluctant, knowing it is not welcome but having to play its part when its cue is called, perhaps trying to say, 'Come, it won't be too bad, don't be afraid, I understand how you feel, but come—there may be other miracles!' No fear of Death, no fight against Death, no enmity toward Death, friendship with Death as with Life. That is—Death for my-self, but not for Johnny, God, not yet. He's too young to miss all the other parts of Life, all the other lovely living parts of Life. All the wonderful, miraculous things to do, to feel, to see, to hear, to touch, to smell, to taste, to experience, to enjoy. What a joy Life is. Why does no one talk of the joy of Life? shout, sing, write of the joy of Life? Looking for books to read with Johnny, and all of them, sad, bitter, full of fear, hate, death, destruction, damnation, or at best resigna-tion. No great books of enjoyment, no sense of great utter simple delight pleasure fun sport joy of Life."

All the things Johnny enjoyed at home and at school, with his friends, with me. All the simple things, the eating, drinking, sleeping, waking up. We cooked, we experimented with variations on pan·cakes, stews, steaks. We gardened, we fished, we sailed. We danced, sang, played. We repaired things, electric wires, garden tools, chopped wood, made fires. We equipped the Chem Lab Workshop, in the made-over old boathouse, with wonderful gadgets, and tried out experiments, both simple and fantastic.

All the books we read. All the lovely old children's books, and boys books, and then the older ones. We read Shaw aloud—how G.B.S. would have enjoyed hearing the delighted laughter of the boys reading parts in *Man and Superman* in the kitchen while I washed up the supper dishes!—and Plato's *Republic* in Richards' *Basic English*, and Russell, and St. Exupéry. On Sundays, we would have church at home: we'd sit outdoors on the beach and read from The Bible of the World, the Old Testament and the New, the Prophets and Jesus, also Buddha, Confucius, and Mahomet. Also Spinoza, Einstein, Whitehead, Jeans, Schroedinger, and Maugham.

We talked about everything, sense and nonsense. We talked about the news and history, especially American history, and its many varied strains; about the roots of his own great double heritage, German and Hebrew; about empires past and present, India's nonviolent fight for freedom, and about reconciliation between Arabs and Jews in Palestine. We talked about Freud and the Oedipus complex, and behavior patterns in people and societies, getting down to local brass tacks. And we also played nonsense games, stunts, and card tricks.

We sailed, and got becalmed, and got tossed out to sea, and had to be rescued. And we planned sailing trips.

All the things we planned! College, and work, and love and marriage, and a good life in a good society.

We always discussed things a little ahead. In a way I was experimenting with Johnny as he dreamed of doing with his elements, as artists do with their natural materials. I was trying to create of him a newer kind of human being: an aware person, without fear, and with love: a sound individual, adequate to life anywhere on earth, and loving life everywhere and always. We would talk about all this as our experiment together.

He did his part in making our experiment a success. Missing him now, I am haunted by my own shortcomings, how often I failed him. I think every parent must have a sense of failure, even of sin, merely in remaining alive after the death of a child. One feels that it is not right to live when one's child has died, that one should somehow have found the way to give one's life to save his life. Failing there, one's failures during his too brief life seem all the harder to bear and forgive. How often I wish I had not sent him away to school when he was still so young that he wanted to remain at home in his own room, with his own things and his own parents. How I wish we had maintained the marriage that created the home he loved so much. How I wish we had been able before he died to fulfill his last heart's desires: the talk with Professor Einstein, the visit to Harvard Yard, the dance with his friend Mary.

These desires seem so simple. How wonderful they would have been to him. All the wonderful things in life are so simple that one is not aware of their wonder until they are beyond touch. Never have I felt the wonder and beauty and joy of life so keenly as now in my grief that Johnny is not here to enjoy them.

Today, when I see parents impatient or tired or bored with their children, I wish I could say to them, But they are alive, think of the wonder of that! They may be a care and a burden, but think, they are alive! You can touch them—what a miracle! You don't have to hold back sudden tears when you see just a headline about the Yale-Harvard game because you know your boy will never see the Yale-Harvard game, never see the house in Paris he was born in, never bring home his girl, and you will not hand down your jewels to his bride and will have no grandchildren to play with and spoil. Your sons and daughters are alive. Think of that—not dead but alive! Exult and sing.

All parents who have lost a child will feel what I mean. Others, luckily, cannot. But I hope they will embrace them with a little added rapture and a keener awareness of joy.

I wish we had loved Johnny more when he was alive. Of course we loved Johnny very much. Johnny knew that. Everybody knew it. Loving Johnny more. What does it mean? What can it mean, now?

Parents all over the earth who lost sons in the war have felt this kind of question, and sought an answer.

To me, it means loving life more, being more aware of life, of one's fellow human beings, of the earth.

It means obliterating, in a curious but real way, the ideas of evil and hate and the enemy, and transmuting them, with the alchemy of suffering, into ideas of clarity and charity.

It means caring more and more about other people, at home and abroad, all over the earth. It means caring more about God.

I hope we can love Johnny more and more till we too die, and leave behind us, as he did, the love of love, the love of life.

FRANCES GUNTHER

Unbeliever's Prayer

———

Almighty God
forgive me for my agnosticism;
For I shall try to keep it gentle, not cynical,
nor a bad influence.

And O!
if Thou art truly in the heavens,
accept my gratitude
for all Thy gifts
and I shall try
to fight the good fight. Amen.

—JOHN GUNTHER, JR.
May, 1946

FINIS

Set in Linotype Baskerville
Format by A. W. Rushmore
Manufactured by The Haddon Craftsmen
Published by HARPER & BROTHERS, *New York*